COGNITIVE INTERVIEWING METHODOLOGY

WILEY SERIES IN SURVEY METHODOLOGY
Established in Part by WALTER A. SHEWHART AND SAMUEL S. WILKS

A complete list of the titles in this series appears at the end of this volume.

COGNITIVE INTERVIEWING METHODOLOGY

Edited by

Kristen Miller
National Center for Health Statistics

Stephanie Willson
National Center for Health Statistics

Valerie Chepp
Hamline University

José-Luis Padilla
University of Granada, Spain

Published by John Wiley & Sons, Inc., Hoboken, New Jersey.
Published simultaneously in Canada.

For general information on our other products and services or for technical support, please contact our Customer Care Department within the United States at (800) 762-2974, outside the United States at (317) 572-3993 or fax (317) 572-4002.

Wiley also publishes its books in a variety of electronic formats. Some content that appears in print may not be available in electronic formats. For more information about Wiley products, visit our web site at www.wiley.com.

Library of Congress Cataloging-in-Publication Data:

Cognitive interviewing methodology / edited by Kristen Miller, National Center for Health Statistics, Stephanie Willson, National Center for Health Statistics, Valerie Chepp, National Center for Health Statistics, Jose-Luis Padilla, University of Granada, Spain.
pages cm
Includes bibliographical references and index.
ISBN 978-1-118-38354-4 (paperback)
1. Interviewing. 2. Cognition. 3. Questionnaires–Methodology. 4. Social surveys–Methodology. 5. Social sciences–Research–Methodology. 6. Psychology–Research–Methodology.
I. Miller, Kristen.
H61.28.C64 2014
001.4′33–dc23

2014011436

10 9 8 7 6 5 4 3 2 1

Dedicated to Janet Harkness,
friend and pioneer of cross-cultural survey methods

CONTENTS

FOREWORD

As an early practitioner of cognitive interviewing, I can remember presenting many talks on this new science throughout the 1990s. Occasionally, an audience member would ask a pointed question: Although its proponents spoke of the cognitive interview as an application of psychology, were we perhaps missing something by not taking into account other disciplines as well—like linguistics, sociology, anthropology, and so on? I thought this to be a good point, despite my strong focus on cognitive psychology as an anchoring point. In fact, over the ensuing years, there have been a number of contributions that have emphasized a wider disciplinary perspective—including the argument that responses to survey questions involve more than just the individual mind of the respondent, especially as they incorporate social and cultural phenomena in a social context.

In the current volume, Kristen Miller and her colleagues provide what I believe to be the clearest statement of this truth, and the furthest point in the evolution of cognitive interviewing as a mature expression of qualitative research that provides a rich multidisciplinary perspective. The arguments, illustrations, and examples within this book challenge practitioners of cognitive interviewing—and more broadly, anyone having an interest in the subtleties of questionnaire design—to think in new ways about how survey questions are developed by designers, answered by respondents, and consumed by data users. In particular, as what I believe to be the main contribution of the volume, they expand our fundamental notion of why we choose to conduct a cognitive interview. Rather than viewing this endeavor only as an attempt to "patch up" deficiencies by identifying and remediating flawed survey questions, the authors conceptualize the cognitive testing enterprise as an opportunity to obtain a more comprehensive view of the items under our microscope. This *interpretivist* viewpoint allows us to alter our underlying research question—so that rather than asking "What's wrong with the survey question?"—we can conversely ask "What's right with it?" More to the point, we can hone that question by asking "How does the question function, and what does this imply about the contexts in which it can profitably be employed?" This expansive viewpoint is clearly of interest across a wide range of applications involving the use of self-report data collection instruments.

Although I use the term "microscopic" above, Miller et al. also further the field of cognitive interviewing by incorporating a vital macroscopic view in leading us to step back and consider the wider context of how survey items function across a range of cultures, languages, countries, and other contexts that are increasingly relevant to survey methodology. The book is the first to tackle the challenges of *comparative* cognitive interviewing, and takes a head-on approach to providing practical assistance to those who face the myriad challenges of question development and evaluation when faced with requirements of instrument translation, interviewing teams that speak different primary languages, and questionnaires that simply do not apply well due to cultural and structural variation. Having collaborated with Dr. Miller in particular over the recent years in which cross-cultural cognitive interviewing has taken root and grown, I can well appreciate the way she has been able to make use of battle-tested experience to save others from having to learn the same hard lessons over again.

A third unique contribution of this volume relates to *analysis*—well-recognized as the Achilles Heel of the cognitive interviewing process. In a word, the authors preach transparency: We need to put our cards on the table in demonstrating exactly what we mean when we say we have conducted cognitive interviews, what our data consist of, and most importantly, how we came to the conclusions we present within a cognitive testing report. Following an increasingly salient thread within the qualitative research tradition, the book provides clear examples, and conceptual direction, concerning how the results of cognitive interviews should be systematically and openly processed, so that a *complete analysis* is conducted. By paying significantly more attention to our analytic processes, we end up with a product that is coherent, defensible, and that sets the stage for replication and further advancement of the field as a whole.

Finally, Miller and colleagues look beyond the cognitive interview to also consider the associated pretesting approaches that exist within our ready toolbox of questionnaire development, evaluation, and testing methods. Although the notion that we can look to alternatives, such as behavior coding, psychometric, and field-based experimental studies, has deep roots in the survey methods field, the current volume advocates tying these roots together, through the use of mixed-method studies that leverage the unique strengths of each approach. In particular, the use of quantitative methods reveals how much, or how often, a phenomenon exists; whereas the overlaying of intensive qualitative methods like the cognitive interview reveals "why this happens" due to the richness of the information the qualitative perspective provides. In summary, the current book provides a clear pathway to new thinking, new methods, and new directions for questionnaire designers, survey managers, and data users.

Gordon Willis

National Cancer Institute

ACKNOWLEDGMENTS

This book has taken us somewhat longer to write than we initially anticipated. The additional time, however, brought additional critique, debate, and refinement of our ideas.

We thank Catherine Simile for providing perspective and significant insight, and Mitch Loeb for his helpful review and input. We thank our colleagues from Swan Solutions, Florencia Ramirez and Luis Cortes, for editorial comments and insurmountable help in pulling together the entire manuscript including figures, tables, bibliography, and appendices. Special thanks go to Lee Burch also of Swan Solutions for his many years of inspiration and support, as well as Karen Whitaker—office manager extraordinaire—who continuously reminds us to think about the "big picture" while keeping us on task in the here and now. We are especially grateful for all our colleagues in the Questionnaire Design Research Laboratory at the National Center for Health Statistics (NCHS) who, collectively, have helped to improve cognitive interviewing methodology.

We also thank the members of the question evaluation community who developed and sharpened the field over the past 20 years. We are particularly grateful for conversation (and sometimes loud debate!) with Gordon Willis, Norman Bradburn, Janet Harkness, Jack Fowler, Paul Beatty, Fred Conrad, Terry DeMaio, Jennifer Rothgeb, Peter Mohler, Rory Fitzgerald, and Debbie Collins—all of whom helped to shape our thinking.

Additionally, we thank our institutions: the National Center for Health Statistics along with the NCHS Office of Research and Methodology which, under the direction of Nat Schenker, promoted and prioritized question evaluation methodology, providing us the resources and time to develop this work. The University of Granada and the Spanish National Statistics Institute, particularly, Miguel Angel Martínez Vidal who pushed the cognitive interviewing projects in Spain.

We are appreciative of Wiley and our editors, Sari Friedman and Steve Quigley, for realizing the value of this project.

A most special thank you to the NCHS Associate Director of Science, Jennifer Madans, who for over a decade pushed us, argued with us, forced us to articulate better (and sometimes drove us crazy!) more than anyone else. Without her mentorship and sincere dedication to question evaluation and the advancement of survey methodology, this book would not exist. For this, we are truly grateful.

CONTRIBUTORS

Isabel Benitez Baena, University of Granada, Spain

Valerie Chepp, Hamline University

Caroline Gray, Research Institute of the Palo Alto Medical Foundation

Meredith Massey, National Center for Health Statistics

Justin Mezetin, National Center for Health Statistics

Kristen Miller, National Center for Health Statistics

José-Luis Padilla, University of Granada, Spain

J. Michael Ryan, The American University in Cairo

Paul Scanlon, National Center for Health Statistics

Alisú Schoua-Glusberg, Research Support Services

Ana Villar, City University London

Gordon Willis, National Cancer Institute

Stephanie Willson, National Center for Health Statistics

1 Introduction

KRISTEN MILLER
National Center for Health Statistics

Although the beginnings of survey research can be traced as far back as the late 1880s, the discussion of question design and the need for standardized questions did not appear for another 50 years (Groves et al. 2009). Since this time, notions about question design have dramatically transformed, particularly in regard to question evaluation. In 1951, Stanley Payne published his book, *The Art of Asking Questions*, and laid out 100 considerations for writing survey questions. Although he maintained that question evaluation studies could be helpful, he argued that the actual writing process should be the higher concern. Today, however, there is a greater emphasis on question evaluation. Also, with the entrance of psychologists, psychometricians, and more recently, anthropologists, qualitative methodologists, and cognitive sociologists, the scientific rigor and scope have increased.

A significant advancement for question evaluation occurred in the 1980s with the entrance of cognitive psychology and the study of the cognitive aspects of survey methodology (CASM). The CASM movement not only brought attention to the issue of measurement error, it also established the idea that individual processes, specifically, respondents' thought processes, must be understood to assess the validity and potential sources of error (Schwarz 2007). The underlying supposition is that, as noted by Willis (2005), "the respondent's cognitive processes drive the survey response, and an understanding of cognition is central to designing questions and to understanding and reducing sources of response error" (p. 23). Thus, with the advent of CASM, the focus of question design shifted from the question writer to the respondent and cognitive processes.

The cognitive processes that make up question response have been represented in a number of theoretical models. A commonly cited question-response model contains four stages: comprehension, retrieval, judgment, and response (Tourangeau 1984; Tourangeau et al. 2000; also see Willis 2005 for a detailed discussion). To provide a response, each respondent proceeds through four specific steps: (1) determining what the question is asking, (2) recalling or retrieving the relevant information, (3) processing the information to formulate an answer, and (4) mapping that answer

Cognitive Interviewing Methodology, First Edition.
Edited by Kristen Miller, Stephanie Willson, Valerie Chepp, and José-Luis Padilla.

onto the provided response categories. By recognizing the cognitive processes, it is possible to understand the complexity of the question-response process as well as the numerous possibilities for response error (Tourangeau et al. 2000; Willis 2004, 2005). By establishing a theoretical foundation for survey question response, the CASM movement provided a basis for scientific inquiry as well as a practical basis for understanding and reducing response error in survey data.

Today there is little debate that question design—how questions are worded and the placement of those questions within the questionnaire—impacts responses (e.g., Fowler 2009; Krosnick and Presser 2010). Newly developed or re-conceptualized methodologies (e.g., latent class analysis, behavior coding, and item-response theory) have repeatedly demonstrated the impact of question design (Madans et al. 2011). Psychometricians, for example, have shown that scale items with more response categories are increasingly likely to produce response distributions with a wider spread than those with fewer categories (Crocker and Algina 2008). Split sample experiments—a method that divides a survey sample into two groups whereupon one group receives a question and the other receives a different version of the same question—have also shown varying estimates (Krosnick 2011; Fowler 2004). In terms of substance and practicality, each methodology has its own benefits but also limitations (see Madans et al. 2011 for in-depth explication). The future of question evaluation lies in the use and integration of varying methodologies. Understanding the range of methodological perspectives—the suppositions, benefits, and limitations—will improve knowledge of question response and survey error, and ultimately ensure quality survey data.

1.1 COGNITIVE INTERVIEWING METHODOLOGY

This book focuses on the question evaluation method of cognitive interviewing—a method arising directly from the CASM movement. It is a qualitative method that examines the question-response process, specifically the processes and considerations used by respondents as they form answers to survey questions. Traditionally the method has been used as a pretest method to identify question-response problems before fielding the full survey. The method is practiced in various ways (Forsythe and Lessler 1991), but is commonly characterized by conducting in-depth interviews with a small, purposive sample of respondents to reveal respondents' cognitive processes. The interview structure consists of respondents first answering a survey question and then providing textual information to reveal how they went about answering the question. That is, cognitive interview respondents are asked to describe how and why they answered the question as they did. Through the interviewing process, various types of question-response problems that would not normally be identified in a traditional survey interview, such as interpretive errors and recall accuracy, are uncovered. DeMaio and Rothgeb (1996) have referred to these types of less evident problems as "silent misunderstandings." When respondents have difficulty forming an answer or provide answers that are not consistent with a question's intent, the question is typically identified as "having problems." A problematic question can then be modified to reduce response error.

By definition, cognitive interviewing studies determine the ways in which respondents interpret questions and apply those questions to their own lives, experiences, and perceptions. In that cognitive interviewing studies identify the content or experiences contained in the respondents' answers, the method is a study of construct validity. That is, the method identifies the phenomena or sets of phenomena that a variable would measure once the survey data is collected. Moreover, cognitive interviewing studies can examine issues of comparability, for example, the accuracy of translations or equivalence across socio-cultural groups (Goerman and Caspar 2010; Willis and Miller 2011). In this way, the method is an examination of bias since it investigates how different groups of respondents may interpret or process questions differently. To this end, cognitive interviewing studies can encompass much more than identifying question problems. Cognitive interviewing studies can determine the way in which questions perform, specifically their interpretive value and the phenomena represented in the resulting statistic.

This book will lay out procedures for conducting cognitive interviewing studies with an eye toward studying constructs, including processes and considerations for data collection, analysis, and making conclusions. The book will also describe how to write results of cognitive interviewing studies so that findings can serve as ample documentation for both survey managers and data users who will use the study to more fully understand and better interpret survey data. Finally, the book will lay out limitations of cognitive interviewing studies and explore the benefits of cognitive interviewing with other methodological approaches. This book is not intended to be a stand-alone guide for conducting a cognitive interviewing study. There are many aspects of the method that cannot be fully addressed in this volume. Other books and articles, such as Willis' (2005) already cited work, *Cognitive Interviewing*, offer significant and complementary material.

Unlike other works, however, the perspective of this book is set specifically within an interpretivist framework in which the construction of meaning is considered elemental to the question-response process. The method explicated in this book, then, is oriented toward the collection and analysis of interpretive patterns and processes that constitute the question-response process. This perspective does not run counter to the psychological focus of cognition, but rather emphasizes interpretive value and the fluidity of meaning within the context of a questionnaire as well as (and perhaps more significantly) within the socio-cultural context of respondents' lives. An interpretivist perspective understands that meanings and thought patterns do not spontaneously occur within the confines of a respondent's mind, but rather those meanings and patterns are inextricably linked to the social world (Berger and Luckman 1966). Context is not identified only as the context of the survey interview, but most significantly it includes the socio-cultural context of that respondent's life circumstance and the perspective that he or she brings to the survey interview. How respondents go about the four cognitive stages—of comprehending, recalling, judging, and responding—is informed by respondents' social location, including such significant factors as their socio-economic status, education, gender, age, and cultural group. Consequently, not all respondents will necessarily process questions in the same manner. An important aspect, therefore, addressed in this book includes a method for examining the socio-cultural influence and comparability across groups.

In thinking about the various objectives that can be accomplished by cognitive interviewing studies, the ultimate goal of a cognitive interviewing study is to better understand question performance. Again, this includes not only identifying respondent difficulties (a.k.a. "problems with questions"), but also identifying the interpretive value of a question and the way in which that question may or may not be consistent with its intent—across particular groups and in different contexts. With a more complete picture of a question's performance, more options emerge in regard to how a question could be altered before fielding or how the resulting variable should be utilized by data users. Moreover, by understanding question performance, a more sophisticated portrayal of response error emerges—one that illustrates response error as a non-binary variable when considered across the entirety of the survey sample. When interpretive findings from cognitive interviewing studies are combined with quantitative studies (as described in Chapter 9), insights into question performance are exponential. A particular limitation of cognitive interviewing methodology is that, while it can discern various patterns of interpretation, it cannot determine the extent to which interpretive patterns exist or vary in the actual survey data. Coupled with a quantitative design, however, it is possible to begin measurement of interpretive variation.

In keeping with the basic tenets of scientific investigation, a predominant theme throughout the book is the necessity for systematic and transparent processes. Systematic data collection and analysis ensure accuracy in the identification of interpretive patterns and processes. Transparency allows readers to understand as well as to cross-examine the ways in which studies were conducted and how conclusions were reached. In addition, transparency instills the trustworthiness of a study and the reputability and believability of study findings. These tenets carry through data collection and analysis to the final report, which must document the analytic process and present evidence to support findings.

The chapters of this book are presented as components of a cognitive interviewing study. Chapter 2 lays out the theoretical foundations of cognitive interviewing methodology, more closely connecting an interpretivist framework to the method that will be articulated in this book. Chapter 3 discusses issues of sampling as well as issues pertaining to quality interview data. The role of the interviewer and the role of the respondent become central themes in the discussion of data quality for cognitive interviews. Chapter 4 lays out a step-by-step process for performing analysis of cognitive interview data while, at the same time, producing an audit trail that links analytic findings with the original interview data. Chapter 5 is a separate analytic chapter devoted to cross-cultural and multi-lingual cognitive interviewing studies. From an interpretive perspective, the impact of socio-cultural context on comparability is a significant component of question evaluation and, therefore, is highlighted in its own chapter. Chapter 6 describes the process for conveying the results of a cognitive interviewing study. In this chapter attention focuses on the importance of transparency and the presentation of empirical evidence—a necessary criterion for producing a credible study. Chapter 7 provides a case study which illustrates how a cognitive interviewing project conducted in the manner presented in this book can be practiced. Chapter 8 presents newly developed tools that benefit cognitive interviewing studies as well as the field of question evaluation. Those tools include Q-Notes, a data entry and analysis application, and Q-Bank, an online resource that, among various other features, houses question evaluation studies and is searchable by question. Chapter 9 discusses limitations of cognitive interviewing studies and illustrates how the method can be integrated with quantitative methodologies to improve understanding of question performance. Finally, the concluding chapter summarizes key principles articulated throughout the book as well as presents emerging ideas and recommendations for the field of question evaluation and survey research.

2 Foundations and New Directions

VALERIE CHEPP
Hamline University

CAROLINE GRAY
Research Institute of the Palo Alto Medical Foundation

2.1 INTRODUCTION

Theory has played a prominent role in the advancement of question design and evaluation. This advancement was ushered in as theories of cognitive psychology were applied to survey methodology. Prior to the advent of the cognitive aspects of survey methodology (CASM) movement, there was little theoretical discussion regarding question response. As Sudman et al. (1996) note, before this time "the work conducted in this domain suffered from a lack of theoretical perspective" (p. 7).

CASM is a critical achievement for survey methodology since theory guides the ways in which empirical research is conducted, as well as why it is conducted in the first place. It also provides insight into why some methods are more appropriate for specific types of research questions than others. Succinctly, CASM established a basis for scientific inquiry into question response and question evaluation. It also laid the foundation for establishing methodological approaches for conducting question evaluation studies.

This chapter will first describe the theoretical perspective underlying the method presented in this book. Specifically, this book is set within an interpretivist framework in which the construction of meaning is considered elemental to the question-response process. The method and methodological considerations presented in this book focus on the collection and analysis of interpretive patterns and processes that constitute the question-response process. This chapter will also describe implications for question response and question evaluation as well as recent directions in the study of interpretation and cognition as it pertains to cognitive interviewing. This discussion focuses on an emerging subfield of interpretivism: cognitive sociology. In addition, three key methodological concepts central to this tradition (narrative, *Verstehen*, and thick description) are examined in relationship to cognitive interviewing methodology.

Cognitive Interviewing Methodology, First Edition.
Edited by Kristen Miller, Stephanie Willson, Valerie Chepp, and José-Luis Padilla.

2.2 SOCIOLOGY AND THE INTERPRETIVIST TRADITION

To date, insights from sociological theory have not been fully integrated into the study of questionnaire design and evaluation. Given that the survey process is fundamentally a social encounter (Sudman and Bradburn 1983; Sudman et al. 1996; Groves and Couper 1998), and given that sociologists have spent the past century theorizing the rules of social interaction (Cooley 1902; Mead 1934; Goffman 1959; Blumer 1969), sociological thought has much to offer the field of question evaluation. In fact, Sudman et al. (1996) argue that "the rules that govern conversations and social encounters in general should help us understand how survey questions are being understood and answered" (p. 1).

A sociological approach to any field of study recognizes that all human behavior takes place in a social environment. The social environment includes many different components, but all sociological work seeks to uncover patterns that arise out of humans interacting with their social worlds. In other words, there is some regularity or "structure" to the social world. Social location refers to the ways in which individuals and groups are differently located within a social structure, based upon socially constructed cultural markers such as race, ethnicity, gender, social class, sexuality, and disability status, among others. Although these cultural markers are socially constructed (i.e., they are not tangible objects but rather meaningful constructs, ideas, or perceptions that are generated through social processes), they have very real consequences in that they intersect to systematically shape individuals' and groups' social experiences as well as their worldviews or interpretations of their experiences (Collins and Chepp 2013). This does not imply that everyone of a particular gender, race, or class interprets the world in the same way, rather social experiences and opportunities that arise out of one's social location shape their interpretations. This includes interpretation of survey questions.

While interpretivist approaches to human cognition have largely been absent from discussions of the method, some researchers have drawn upon interpretivist sensibilities in their conceptualizations of cognitive interviewing (Gerber and Wellens 1997; Gerber 1999; Miller 2011). For example, Gerber (1999) argues that the utility of incorporating interpretivist modes of analysis—and the focus is specifically on ethnography—into the survey context pivots on its "ability to represent complexity" (Gerber 1999, p. 219; see also Miller 2011).

Although sociologists are generally interested in identifying patterns of human interaction with the social world, like any discipline, sociology has numerous subfields, each rooted in different intellectual traditions. Interpretivism emphasizes the meaningful quality of individuals' engagement in the social world. Furthermore, it recognizes that understanding of the social world is filtered through a complex set of interpretations that are variously informed by social experiences and cultural contexts.

Interpretivist approaches attempt to not only identify but also to understand the different realities that social actors construct. For instance, in the case of a survey question that asks respondents how many times they visited a doctor in the past year, there are numerous potential interpretations that this question might elicit that may be shaped by social factors such as a respondent's age, education level, cultural background, health insurance status, or health condition. Some respondents, for example, may interpret the term "doctor" to mean general practitioner, while others may include specialists such as surgeons, gynecologists, dermatologists, allergists, or podiatrists. Still others may understand the term more broadly to include visits to the dentist or eye doctor. In addition to conventional definitions of "doctor" rooted in a Western medical model, respondents might variously include (or not include) visits to non-traditional practitioners, such as midwives, chiropractors, or acupuncturists.

Although some of these interpretations may not be the intent of the question designer, none of these interpretations are inherently "wrong." It is understandable that interpretations may vary across respondents given their different circumstances and experiences. Moreover, individual respondents' interpretations could shift if their circumstances or experiences change. Among interpretivists, it is widely accepted that multiple and fluid meaning patterns can exist and shift over time. It is the analyst's responsibility to identify and make sense of these varying and potentially shifting interpretations across respondents. As a result of this analytic work, it is possible to identify how a question performs across a range of respondents with differing backgrounds or experiences. Within a framework that emphasizes meaning as an elemental component of question response, a cognitive interviewing study can detail the various phenomena captured by a question and, ultimately, represented in a survey statistic.

2.3 NEW DIRECTIONS: INTERPRETATION AND COGNITION

Although the four stages of question response—comprehension, retrieval, judgment, and response—have traditionally been understood as uniquely cognitive psychological processes, cognitive psychologists are not the only scholars interested in mental processes. Cognitive sociology, a relatively recent theoretical and empirical development in sociology, is a subfield of interpretivism that provides particularly fertile ground for thinking about cognitive interviewing methodology within an interpretivist framework.

The aim of cognitive sociology is to demonstrate the numerous ways in which cognitive processes can be understood from a sociological perspective, suggesting, above all, that cognitive processes are shaped by cultural phenomena. The field is interested in understanding the categories, schemes, and codes that individuals use to organize their thoughts and make sense of the world around them. Rather than view these categories, schemes, and codes as universal as many cognitive psychologists do, cognitive sociology instead argues that these "thought structures" are the product of the social environment (Zerubavel 1997). However, although such thought structures are not universal and static, they are not idiosyncratic or individual either. The thoughts that enter minds are rooted in broader social processes and social relations. The most obvious example of this is language itself. Languages are only meaningful insofar as members of a language community share that same language and can understand what the different words it encompasses mean. But language can be understood more figuratively as shared meaning systems where individuals become socialized into shared (though not always) ideas and thought patterns. Members of society become socialized into distinctive "thought communities" (Zerubavel 1997) that both inform and reflect the inter-workings of their minds.

Cognitive sociology is interested in many of the same processes as cognitive psychology, but analyzes these cognitive processes in a different way. For example, DiMaggio (1997) outlines how cultural mental schemata inform "the way we attend to, interpret, remember, and respond emotionally to the information we encounter and possess ... Culture inheres not in the information, nor in the schemata, nor in the symbolic universe, but in the interactions among them" (p. 274). Brekhus (2007) outlines six cognitive processes that the field of cognitive sociology regularly examines from a cultural perspective; these are perception, attention/inattention, memory/time/chronology, classification, meaning making, and social identity. These processes articulate with and build upon the cognitive processes traditionally conceptualized as the question-response model. Although the processes outlined by Brekhus (2007) are not identical to those discussed by Tourangeau (1984), a comparison of the two models illustrates how cognitive processes deemed important to the question-response process share similarities with those processes that cognitive sociologists have empirically shown to be shaped by culture.

Specifically, *perception* is closely tied to comprehension. Perception refers to the various ways people perceive the world which, cognitive sociologists argue, is filtered through shared understandings of the world with the people around them (DeGloma and Friedman 2005). *Attention and inattention* speaks to issues of comprehension and judgment. Here, cognitive sociologists draw upon Erving Goffman's (1974) work on frame analysis to show how individuals place mental frames around their social reality, deeming most things unworthy of attention. This work is useful for illustrating the ways in which survey respondents variously pay attention to different social experiences when answering survey questions. Seasoned cognitive interviewers are familiar with respondents' claims to have "forgotten" or "not think" about relevant life experiences when answering survey questions. A better understanding of how social actors direct their attention can help cognitive interviewers gain a better understanding of how respondents comprehend and respond to survey questions.

The role of *memory, time, and chronology* in the question-response process is captured in the retrieval stage. Tourangeau (1984) emphasizes the types of information that the respondent must excavate in order to accurately answer the question. Obviously this depends on the respondent's memory. But memory is also a social process. What individuals remember and do not remember, and relatedly, *how* they remember an event is shaped by the social environment. Sociologists studying "collective memory" have illustrated how historical accounts are variously interpreted and thus how the cognitive process of memory is intimately tied to past and present social relations (Olick et al. 2011).

Exchanging meaningful symbols involves elaborate *classification* schemes. Though the desire to classify is universal, the classification schemes that are produced are always culturally mediated (Durkheim 1912; Levi-Strauss 1962). Interpretivist approaches are particularly apt at describing the different types of classifications that societies, cultures, and small groups use to organize things. This observation—that human beings have a fundamental desire to classify—is particularly relevant to explaining the question-response process because the questionnaire designer must be able to accurately capture and articulate the same types of classification schemes that respondents themselves use. Uncovering these schemes and categories is the task of the cultural analyst and also the cognitive interviewer. It becomes clear in a cognitive interview when the respondent's classification categories are counter to those offered by a question's response options.

While cognitive sociological and other culturally informed perspectives contend that thought patterns are not universal and unchanging, one aspect of the human mind that they assume is in fact universal is the need to engage in the meaningful exchange of symbols, or what is commonly thought of as language. Out of language emerges meanings or *meaning making*, a fifth cognitive process outlined by Brekhus (2007), which deeply shapes how respondents comprehend, retrieve information, and judge how to respond to survey questions. Finally, cognitive sociologists are interested in issues of *social identity*, which relates to the previous discussion about social location. Cognitive sociologists argue that, as a result of their social location in particular cultures and sub-cultures, individuals are socially located in particular thought communities (Brekhus 2007: 450). For example, Brekhus (2003) demonstrates how the social location of gay men suburbanites shapes their understandings of stigma and privilege.

2.4 METHODOLOGICAL IMPLICATIONS FOR COGNITIVE INTERVIEWING

Within sociology and the social sciences more broadly, different theoretical traditions tend to cluster around different social scientific methods. Typically, interpretivist theories map onto qualitative methods that seek to account for cultural variation, human agency, and historical contingency. In the social sciences, qualitative methods are distinguished by their emphasis on meaning making among cultural groups, while quantitative methods draw aggregate conclusions about a population. Qualitative approaches to capturing meanings, however, have gone through various iterations. For example, Emerson (2001) recounts how, in ethnographic field research of the nineteenth century, primary data from field observations and interviews were not collected by scholar experts. Instead, this task was performed by non-scholars who already had some familiarity with a region or culture, such as missionaries, traders, and explorers. The prevailing belief among scholars at this time was that all the important intellectual work took place after the field data were collected, back in the scholar's "home" country. As such, it was believed that anyone could collect data, including untrained non-professionals. However, concerns over data quality led turn-of-the-century scholars to collaborate more closely with those working in the field to improve data collection techniques. This included academically trained scholars overseeing field research teams, and the insistence that fieldworkers be fluent in the native language. Nonetheless, despite efforts to better capture local meanings, many ethnographers continued to privilege theorizing over data collection and fieldwork. The fieldworker's role took on another dimension around World War I, during which time fieldworkers began to acknowledge the cultural biases inevitably present in their data collection techniques, including how they constructed their interview protocols and survey questionnaires; as such, fieldworkers began to rely on methods of direct observation and later participant observation in addition to interviews and questionnaires (Emerson 2001).

This move toward observation was associated with a move toward the recognition of the important role that the data collector plays in qualitative research. No longer could the data collector be an untrained professional, or even a detached, objective observer. Rather, the knowledge and expertise of the data collector was an integral component to the quality of data collected, as the data collector had to be theoretically and methodologically well-versed enough to interpret and analyze data as it was being collected, in order to inform subsequent decisions and approaches to field data collection. This historical perspective of ethnography is useful for understanding an interpretivist approach to cognitive interviewing in that it recognizes the important role of the data collector in the qualitative research endeavor. As with all qualitative methods, the methodological training and expertise of the cognitive interviewer is essential to the data that is collected and subsequently analyzed in cognitive interviewing studies (see Chapter 3).

Today, interpretivist approaches to qualitative research encompass a wide range of methodological techniques including observation, interviews, action research, close readings, focus groups, discourse analysis, comparative case studies, historiographies, personal narratives, auto-ethnography, and visual analysis, among many others. Across these qualitative methods, researchers have developed various methodological concepts to help them specify what it is they do in their interpretivist analyses of empirical work. Below are three concepts that are particularly useful for understanding an interpretivist approach to cognitive interviewing: narrative, *Verstehen*, and thick description.

Narrative. Narrative, simply put, is a way of organizing and making sense of experiences by putting those experiences into a structured sequence of events with a beginning, middle, and end (White 1987). The concept of narrative is closely associated with interpretation. In fact, the term narrative sociology has been used interchangeably with interpretive sociology (Reed 1989); in this sense, in the context of sociological work, narrative is understood as work that is rhetorically descriptive, sequential, and analytically interpretive. However, the term narrative can also refer to something more specific. Social scientists have increasingly recognized that individuals understand themselves and their life histories and circumstances in the narrative form. Narrative has been used not only as a concept for elucidating social processes but also as a methodological tool meant to aid analysis (Richardson 1990; Franzosi 1998). This is especially true in qualitative research.

An interpretivist approach to cognitive interviewing research applies to conceptions of narrative in that narrative is one of two cognitive modes by which individuals construct and order meaning (Bruner 1986). In this way, during the cognitive interview, a researcher should be especially attentive to a respondent's narrative as it provides key insight into the ways the respondent makes meaning of the survey question (Miller 2011). Analytic efforts to make sense of narrative data are commonly referred to as "narrative analysis" (Gotham and Staples 1996; Franzosi 1998). This larger methodological tradition of attending to narrative, particularly in qualitative research, is very useful for interpretivist approaches to cognitive interviewing. Documenting and analyzing the stories people tell about how they answered a survey question is a powerful approach to improving questionnaire design. Yet, in addition to collecting respondents' narratives as an important cognitive interviewing technique, narrative can also be thought of as a "culture structure," in that it enables respondents to make sense of their experiences (Alexander and Smith 2004). Similar to economic, political, and organizational forms, culture is also highly structured and functions as a broader social context that shapes individuals' actions and interpretations. In this way, narrative is useful for understanding cognitive processes because it provides a structure for sense-making, and it is this broader context or "structure" that researchers essentially aim to uncover in the cognitive interview.

Verstehen. Rooted in nineteenth century German sociology, the concept of *Verstehen* refers to the qualitative methodological enterprise by which the researcher aims to understand a group on their own terms. The concept later became closely associated with the methodological orientation of twentieth century German sociologist Max Weber, who conceived of *Verstehen* broadly as an interpretive understanding, and specifically "adopted the idea that the 'understanding' (*Verstehen*) of meaning is essential to the explication of human action" (Giddens 2001: xi).

As cognitive interviewers working in an interpretivist tradition, respondent narratives are taken at face value. Indeed, as Dilthey observes, individuals only know one particular thing: their own lives (Alexander 1987: 186). This is why respondents' interpretations are taken at face value. Even if a respondent "misinterprets" a survey question (i.e., interprets a question other than how survey researchers intend), their interpretation is nonetheless "right" and useful from an analytic perspective. It is for this reason that "expert review" of survey questions is not an advocated evaluation approach. Nor should respondents be asked to speak for others, or asked to hypothesize other scenarios, such as what they would do if the questions were worded differently or if their circumstances were different (for more discussions on this, see Chapter 3). Rather, an interpretivist approach to cognitive interviewing aims to recruit respondents from different backgrounds in an effort to undercover (all) patterns of interpretation and saturate categories. Drawing upon principles of *Verstehen* and through probing, it is possible to elicit a narrative that can shed light on why a respondent interpreted a question in a certain way.

Thick description is another concept that is central to interpretivist theories and methods (Geertz 1973). Its aim is to collect rich, thickly detailed accounts of some aspects of social life. As Denzin (1994) argues, thickly described accounts "give the context of an experience, states the intentions and meanings that organized the experience, and reveals the experience as a process. Out of this process arises a text's claims for truth" (p. 505). In general, qualitative researchers hold rich, thickly described scenarios as ideal data that can later be analyzed.

In the cognitive interview, the aim is similar: to capture a fully and richly described narrative of how the respondent answered an interview question. As will be more fully described in the next chapter, this narrative can then be transcribed verbatim or video recorded, and then thickly described and synthesized into a detailed summary. Summaries should include "observer comments" or, reflections, hunches, and thoughts taking place in the researcher's head throughout the data collection and analysis process (Merriam 2009). A thickly described cognitive interviewing narrative is important because, as Geertz (1973) argues, meanings are multilayered, and simply describing an interaction at the surface may not fully extract the "true" meaning of the situation.

In regard to cognitive interviewing, the interviewer should thickly describe a respondent's answer to a question—did the respondent express uncertainty in their answer, did they count on their fingers to arrive at an answer, or did they change their answer from their initial response? Further, the researcher should thickly describe the probing process. For example, are there inconsistencies in the respondent's narrative? If answering a sensitive question about stigmatized behavior or illegal activity, does the respondent appear nervous? Does the researcher believe the respondent is telling the truth? Through a thickly described account, the cognitive interviewer can capture a fuller account of the meaning of the situation. In this way, cognitive interviewing is "detective work," not only in the sense suggested by Willis (1994) in that cognitive interviewers look for clues about questionnaire problems, but more so because cognitive interviewers aim to get at underlying meanings that represent the foundation of the question-response process.

2.5 CONCLUSION

This chapter outlined the interpretivist approach that serves as the theoretical framework for the method described in this book. Interpretivist perspectives have much to offer the field of cognitive interviewing methodology, as they can help to inform and improve data quality and survey instruments by focusing on interpretative value of the various phases of the research process. Newly emerging fields of interpretivist research such as cognitive sociology lend credence to an interpretivist approach to cognitive interviewing, as scholars have theoretically and empirically demonstrated that cognitive processes, which have long been of interest to cognitive interviewing practitioners, are fundamentally shaped by cultural phenomena.

3 Data Collection

STEPHANIE WILLSON and KRISTEN MILLER
National Center for Health Statistics

3.1 INTRODUCTION

Cognitive interviewing studies can achieve multiple research goals regarding the performance of a survey question. The types of analytic conclusions that a study can ultimately make stem from methodological decisions determined throughout the study, beginning with data collection. In a cognitive interviewing study, two topics are pertinent to data collection: sample selection and the interview itself. This chapter reviews sample design issues, including sample composition and sample size. In discussing the interview process, the chapter will address the role of the interviewer and the role of the respondent. It is necessary to consider what respondents can and cannot know about their experiences or assessments of a question as well as how they can best convey that information. The interviewer is responsible for both recognizing and eliciting the kind of information that is knowable to the respondent. A central theme emerges throughout this chapter: data quality. In this discussion, the characteristics of data quality for cognitive interviews are addressed, as well as how best to achieve those criteria. Taken together, these key components of data collection impact the quality of a cognitive interviewing study and the extent to which study conclusions can be made.

3.2 COGNITIVE INTERVIEWING STUDY SAMPLE

When beginning a cognitive interviewing study, decisions must be made about respondents, specifically, who and how many people should be interviewed. These are the most common questions among practitioners of cognitive interviewing studies. They are also the most difficult questions to answer because there is no single correct answer. Sample size and characteristics are dependent on the complexity of the

Cognitive Interviewing Methodology, First Edition.
Edited by Kristen Miller, Stephanie Willson, Valerie Chepp, and José-Luis Padilla.

questions that are being evaluated, as well as the ultimate goal of the study. Sample decisions are also dependent upon the ongoing analysis of interviews and, therefore, can rarely be predetermined at the outset of a cognitive interviewing study.

As is the case for any scientific study, the sample design stems from the specific goal of a study. Cognitive interviewing studies, for example, differ dramatically in purpose and, therefore, in design from survey research. The goal of survey research is to collect discrete, quantitative data from a portion of a population so that estimates can be generalized to a larger population. The primary objective of cognitive interviewing, however, is not to make use of statistical inference. Rather, the objective is to collect in-depth, thematic understandings of patterns and processes utilized by respondents to answer survey questions. While survey research employs a deductive, quantitative methodology and relies on a relatively large random sample to achieve statistical inference, cognitive interviewing employs an inductive, qualitative methodology and, consequently, draws upon a relatively small, non-probability sample.

In looking to build thematic schemes and conceptual understandings of question performance, it is necessary to logically determine and identify respondents that will most likely experience or process questions in a variety of ways. This is quite different from building a sample in which every individual has a known probability of being included. To build a sample based on probability or chance means that many more interviews will be required than is necessary or even helpful. In short, rather than drawing upon a random sample, a cognitive interviewing study employs a purposive sample, one that is theoretically driven and is deliberately selected to achieve a particular goal (Lincoln and Guba 1985). For example, if the goal of a cognitive interview study is to evaluate questions on smoking behavior, the sample design would include people who smoke. A probability sample is not ideal to the extent that it might produce a sample with no smokers at all.

At this point, a general framework of cognitive interview sampling has been described. There is little, if any, disagreement among those in the field regarding these principles. There are, however, different ideas regarding specific decisions for attaining the ideal number and type of respondents. For the most part, the different sampling approaches stem from the various conceptualizations of cognitive interviewing, as well as its conceptualized purpose.

3.2.1 Considerations of Sample Design

Traditionally, cognitive interviewing studies have been guided by the goal of finding problems in survey questions (Beatty and Willis 2007; DeMaio et al. 1993; Willis 2005). The ideal sample size, then, is dictated by the number of interviews required to assure the researcher that all of the problems have been identified. With this goal, as Willis (2005) has argued, respondents are chosen to obtain the "greatest cross-section of the population as possible in order to identify a wide range of problems" (p. 140). The best samples, then, contain the diverse characteristics of respondents represented in the survey sample. In addition, respondents are chosen according to various skip patterns and topics in the questionnaire. For example, if the questionnaire asks questions of both smokers and non-smokers, then the cognitive interviewing study should include both types of respondents.

One difficulty with this type of sampling, however, is in determining how much demographic diversity is sufficient. Is it important for the cognitive interview sample to be composed of half men and half women? Is the racial composition and educational status of the sample important? What about the health status? Those who live in rural areas versus metropolitan areas? Given the small sample size of a cognitive interviewing study, it is impossible to attain a sample that would contain all characteristics of a survey sample. Constructing a cognitive interview sample solely on the basis of demographic diversity can lack a compelling justification for why particular groups are chosen and others are not. As would be the case for random sampling, this type of sampling could also result in an unnecessary number of interviews being conducted.

Another direction for determining sample composition is not necessarily to select a diverse group of respondents, but rather to choose those who are more likely to reveal the most problems. Ackermann and Blair (2006), for example, have argued that respondents with higher education make the best respondents because they can more easily identify and articulate the most number of problems within a single interview. A similar argument could be made for respondents with less formal education, who are more likely to illuminate problems such as difficult vocabulary or questions with vague intent.

When sampling in this manner, large sample sizes are unnecessary because problems will (or should) be revealed immediately. Sheatsley (1983), for example, states that "it usually takes no more than 12–25 cases to reveal the major difficulties and weaknesses in a pretest questionnaire" (p. 226). Sudman (1983) similarly advocates "20–50 cases is usually sufficient to discover the major flaws in a questionnaire" (p. 181). Although Willis (2005) advises that interviews should be conducted until all the major problems have been detected and satisfactorily addressed, he also points out that, in reality, sample size is often dictated by the survey schedule and notes that 5–15 is a typical range. Others, however, argue that larger sample sizes are necessary to give researchers more opportunity to uncover problems. Indeed, Blair et al. (2006), in another study, note that they continued to find problems with a sample as large as 50 cognitive interviews, discovering additional problems late in the interviewing process.

A persistent difficulty faced by researchers using any of the above strategies remains knowing exactly how many interviews to conduct. Because the intent of the study is to identify problems, the primary question becomes: "how do I know when I have found all of the problems associated with a particular survey question?" In some instances, researchers may ask: "How do I know when I have found the most significant problems?" For those attempting to capture all potentially relevant characteristics in a sample, the problem becomes knowing how many interviews to include for each respondent characteristic type.

Although these approaches have attempted to establish much needed guidelines regarding the sample size, in the end they do not fully address these issues and cannot necessarily apply to all types of cognitive interviewing studies. For example, cognitive interviewing studies that examine the performance of demographic questions such as age or gender will likely require fewer interviews than questions about household income or satisfaction with the current political administration.

In addressing the guidelines for sample size and composition, it is helpful to turn to qualitative methodology literature, specifically, the grounded theory literature. Based on the seminal work of Glaser and Strauss (1967), a grounded theory approach uses *theoretical saturation* to address questions of sample size and *theoretical relevance* for questions of sample composition. Within this framework, sampling design is contingent upon analysis of the interviews and the two (sampling and analysis) must ultimately occur simultaneously.

3.2.1.1 Theoretical Saturation For any qualitative study, the purpose of data collection is to collect pertinent data that can elaborate and refine understanding of a particular phenomenon. Within the context of cognitive interviewing, the purpose is to gather information pertaining to question performance, specifically, to understand how and why potential survey respondents answer questions the way they do. To understand question performance in its complete form, cognitive interviewing studies should: (1) identify any difficulties that respondents experience when answering a question, (2) identify the constructs that respondents consider and include in their answer and, (3) identify if and why a particular group of respondents process a question differently. For grounded theory studies, interviewing must continue until the entire phenomenon of study is explained. When this criteria is met, theoretical understanding is said to be "saturated" (Charmaz 2006; Glaser and Strauss 1967), also known as reaching "theoretical sufficiency" (Dey 1999). Within the context of a cognitive interviewing study, theoretical saturation occurs when the full range of problems, interpretations, and constructs are identified and explained. If an interpretation demonstrated by a respondent cannot be explained within the context of other interviews, then more data are needed.

An example from a question evaluation project conducted at the National Center for Health Statistics demonstrates theoretical saturation within a cognitive interviewing study. The study examined questions for the 2007 complementary and alternative medicine (CAM) component of the National Health Interview Survey (Willson 2006). The survey questions were designed to measure the use of CAM practices that are both practitioner-based (e.g., acupuncture) and self-practiced (e.g., use of herbal supplements). The questionnaire included an expanded section on the use of herbal supplements and vitamins. This section, in particular, proved to be problematic.

The lead-in question to the herbal supplement section was designed to filter respondents into or out of the section. It asked, "Have you ever used natural herbs or other non-vitamin/non-mineral dietary supplements for your own health or treatment?" The question elicited numerous false positive and false negative reports. That is, many respondents did not respond according to the intent of the question, misreporting the times when they did or did not use herbal supplements. Not unexpectedly, the analysis of the first cognitive interviews revealed that respondents who did not understand the term "natural herb" tended to respond erroneously. However, analysis also showed that some respondents who *did* understand the term as intended also responded erroneously. The response problems, therefore, could not be explained as a simple case of problematic vocabulary.

With further analysis it was discovered that erroneous reports could also be attributed to respondents' perceptions of self, particularly in regard to their own health status and health habits. That is, some respondents identified as belonging to a certain lifestyle associated with alternative medicine. Not surprisingly, these respondents easily envisioned themselves as taking supplements even if they did not. Others—though using herbal supplements at the time—did not identify with the lifestyle, sometimes even disassociating from this type of persona. Instead of interpreting the question as a simple question about the use of herbal supplements, many understood the question as asking whether they identified with an alternative medicine lifestyle. It was this phenomena regarding respondents' self-perception—not only a vocabulary problem—that contributed to erroneous reports.

In order to fully understand these patterns, it was necessary to include the range of respondent conceptualizations: those who do and who do not align themselves with this persona or lifestyle. In addition, because most respondents in the initial sample used herbal supplements, interviews with respondents who did not use herbal supplements were required. These additional interviews provided the ability to fully explore the interpretive processes leading to false negative as well as the false positive reports.

In the context of studying meaning, sample size should be dictated by a full explication (i.e., saturation) of thematic or interpretive patterns that explain the question-response process. By fully identifying interpretive patterns, it is possible to understand why respondents go about answering questions the way that they do—and why different respondents may use different processes to go about answering those questions. The ultimate number of interviews conducted in any one particular cognitive interviewing project, then, is based not on a particular numerical goal, but on the ability to construct a theory that explains why respondents answer questions the way they do and, in the end, the construct (or set of constructs) that a particular question captures.

3.2.1.2 Theoretical Relevance The criterion governing the composition of a sample is closely tied to that of the sample size. Beyond demographic diversity and characteristics associated with the question topic (e.g., smokers and non-smokers), there is little guidance in knowing when enough sample diversity is attained. When following a grounded theory approach, diversity is also important, but is more directed and defined. Respondents (or groups of respondents) are chosen on the basis of theoretical relevance to emerging patterns. Glaser and Strauss (1967), for example, note that "the researcher chooses any groups that will help generate, to the fullest extent, as many properties of the categories as possible and that will help relate categories to each other and to their properties" (p. 49). For a cognitive interviewing study, sample characteristics should be chosen to elicit the greatest diversity in regard to the various processes or patterns that respondents may use when answering a survey question. Demographic diversity and diversity in characteristics related to the question topic are good places to begin sample selection. But as patterns of interpretation and processes are captured, sampling should shift to help develop a complete understanding of question performance. Typically, this is achieved by comparing groups that are similar to and different from each other. In light of the example presented above, once the concept of "alternative health identity" emerged, respondents were chosen based on their use of herbal supplements and on their self-conceptions regarding the practices of alternative health.

A well-considered and ongoing sampling plan is a necessary component of a cognitive interviewing study that will be able to provide a complete portrayal of a question's performance. Once a sampling plan has been developed, an interviewing plan is also required. The next section addresses data collection within a cognitive interview.

3.3 THE COGNITIVE INTERVIEW

The previous section discussed how sampling decisions impact the types of conclusions that can be made by a cognitive interviewing study. The next section discusses how the interview process itself also shapes the types of conclusions that can be made. Whether explicitly addressed or not, researchers make assumptions throughout the cognitive interview, including assumptions about what respondents know, how they know it, and how they can communicate it. This is an integral but under-acknowledged aspect of cognitive interviewing methodology. In an effort to improve the overall quality of cognitive interviewing studies, it is important to recognize and make explicit these underlying assumptions. In considering what respondents are capable and not capable of reporting, it is possible to determine what interview data is relevant and of higher quality.

3.3.1 Differing Approaches to Cognitive Interviewing

The practice of cognitive interviewing has grown out of a tradition of cognitive psychology that emphasized a need to understand the mental reasoning and processing that occur in the mind of a respondent answering a survey question. As a pretest method, the goal of cognitive interviewing studies is to find and correct any problem (i.e., respondent difficulty when answering a question) associated with a particular question. The purpose of the cognitive interview, then, is to extract information about respondents' cognitive processes in order to determine where difficulties arise. Various approaches have emerged regarding the best way to elicit respondents' cognitive processes.

Early techniques of cognitive interviewing were dominated by the *think-aloud* as a method to collect data on the cognitive aspects of the question-response process. With this approach, respondents verbalized their thought processes as they went about answering a survey question. This allowed the interviewer to determine and document respondents' mental processes. Often, the interviewer would attempt to be as neutral and uninvolved as possible so that the undirected thoughts of respondents could be captured without bias. In fact, some advocates of this approach impressed the importance of the interviewer being as absent from the interaction as possible, for example never referring to themselves, omitting the word "me" from any language or questions asked by the interviewer, and even remaining out of the respondent's vision (Ericsson and Simon 1980; 1993). In addition, data quality was tied to the type of memory used by the respondent; short-term as opposed to long-term memory was deemed of higher quality in respondents' reports of their cognitive processes. Hence, the technique was applied using concurrent verbal reporting, that is, during question administration and not after, so the interviewer could be sure the respondent actually remembered what they were thinking and did not fabricate their thought process after the fact.

Gradually, the use of the think-aloud faded in favor of *verbal probing* (Willis 2005); however, the primary goal of capturing psychological thought processes remained. The primary difference between the think-aloud approach and verbal probing is that, with verbal probing, the interviewer becomes present and more active in the interview. Rather than being as unobtrusive as possible, the interviewer "probes" the respondent with direct questions about their thought processes during the question-response process, which can occur either concurrently or retrospectively. The advantage of this approach is that the interviewer is able to collect specific data regarding the four-stage cognitive model (Torangeau et al. 2000). Willis (2005) identifies typical probes designed to capture each stage of question response including "What does this term mean to you?" (comprehension), "How do you remember that?" (recall), "How did you arrive at your answer?" (judgment) and "Was it easy or difficult to choose an answer?" (response). Verbal probing is also thought to be more advantageous for certain types of survey questions. For example, Wilson et al. (1996) found that think-aloud interviews do not work well for questions that ask for a good deal of self-reflection, such as those asking respondents why they do something or those that ask for an attitude or opinion. Respondents tend not to (or cannot) spontaneously report that level of detail about their cognitive processes. Instead, verbal probing was defined as a more appropriate way to obtain this information as interviewers are able to guide respondents through their cognitive processes.

An *interpretivist approach*, on the other hand, moves away from the goal of understanding cognitive processes independent from a socio-cultural context. Instead, this approach focuses on respondents' interpretations and, specifically, how their own lived experiences inform their answers to survey questions. Gerber and Wellens (1997), for example, suggest that the primary goal of these types of cognitive interviewing projects—which are increasingly conducted in the federal statistical system—is to understand the meanings people bring to survey questions, not to understand cognitive processes, per se. For example, Miller (2011) notes that "the interview text is not the essence of respondents' cognitive processes; at the very most, it is respondents' interpretations of their cognitive processes" (p. 60). In outlining a more interpretivist approach, Miller describes the interview as collecting a story—or "narrative"—that details how and why respondents answer questions the way they do. Beatty (2004) also acknowledges a trend in this direction. He suggests that "much interviewer probing seems more oriented toward generating more explanations of respondent circumstances and how those circumstances fit the parameters of the question" (p. 63). Hence, verbal probes are still used, but in a different way. Miller (2011), for example, argues that since the goal of the interviewer is to capture as much of the narrative as possible, it is important for the interviewer to ask whatever questions are necessary to fill in gaps or to address apparent contradictions.

It is important to consider the epistemological assumptions rooted within various approaches to cognitive interviewing, as this shapes how we make sense of the data that is collected. Specifically, what do the various approaches assume regarding what can be known and who is the knower? Both think-aloud and verbal probing approaches, for example, assume that respondents can report and even evaluate their own cognitive processes. This assumption—that respondents can provide accurate reports of their own cognitive processes—forms the logical basis for think-aloud interviewing. Advocates of the verbal probing approach, however, argue that many cognitive interviewing respondents cannot speak to these processes and that the think-aloud is a difficult task for many respondents to perform. This was a primary reason that the pure think-aloud technique was abandoned by many practitioners.

An interpretivist approach to cognitive interviewing assumes that respondents understand and process survey questions through their own personal experience and can explicate this understanding best through the narrative process. In addition, an interpretivist approach assumes that respondents can only be informants to their own experience. It is the role of the analyst to makes sense of the entirety of data from all respondents' perspectives. This approach does not assume that respondents can always report their cognitive processes, specifically, the four stages of the question-response process—regardless of whether or not they are outright asked about them. The interpretivist approach also assumes that the respondent is not the best judge of the quality of a survey question, nor that they can even determine whether or not they misunderstood a particular question. In short, the respondent is not a question design expert and should not be placed in this role. This approach posits that it is only through analysis of narratives—by examining contradictions, incongruencies, patterns of interpretation in respondents' stories—that response errors may be understood and explained. Significantly, this approach most succinctly identifies the actual phenomena that respondents include in their answer—the construct that is ultimately captured by the survey question. In keeping with the goal of achieving narrative data, then, experience-based probes are used. Rather than asking respondents to be question design experts, experience-based probes assume only that respondents are "experts" on their own personal experiences, that is, they can only serve as informants to their own experiences. As such, these types of probes are follow-up questions which ask respondents to *describe and explain* this experience. This yields data that convey authenticity and convinces us that we understand why respondents answered the survey questions the way they did.

3.3.2 Different Kinds of Data: Respondent as Evaluator versus Respondent as Story Teller

When respondents are interviewed, they are necessarily placed in the position of knowledgeable informant. The type of informant, however, is typically determined by the type of probe questions that they are asked. In some cases, the cognitive interviewing respondent is asked to evaluate parts of the question (e.g., the response categories), the question itself, the overall questionnaire, or their own cognitive processes as it relates to the success of the question. In instances where respondents are placed in the role of an evaluator, respondents can often fall into the role of questionnaire designer—a role that, with few exceptions, is inappropriate for them to assume. If the goal of the study is to collect respondent interpretations, probes that place the respondent in the role of evaluator typically result in poor data quality that reveal little information regarding the actual response process.

The following examples of interviews illustrate how interviewers' probes place respondents in an evaluator role. Example 3.1 illustrates how probes that ask for opinions can position a respondent in an evaluator role, which tends to generate an interview text that does not shed light on thought processes. The specific probes are identified by *italics*.

EXAMPLE 3.1 RESPONDENT AS EVALUATOR

Interviewer: Do you have a physical or mental health condition or a problem with memory, pain, or fatigue that has required you to reduce or change regular activities you want or need to do? Not at all, some of the time, most of the time, all of the time?

Respondent: Some of the time.

Interviewer: Some of the time? *And what did you think of that question?*

Respondent: I was interested in that question. Could you read it again?

Interviewer: Sure. *Is it too much at one time?*

Respondent: Umm ... possibly, but uh, what I was interested in is what came first. The sequence of the things I was to think about.

Interviewer: Okay. I will read it again. Do you have a physical or mental health condition or a problem with memory, pain, or fatigue that has required you to reduce or change regular activities you want or need to do?

Respondent: No. I like it. It stated things differently.

Interviewer: *And you like that?*

Respondent: Yeah.

Interviewer: *And it is not too complicated?*

Respondent: No, it is not too complicated.

By asking the respondent what she thought of the question (as opposed to why she answered as she did), little information is collected regarding her thought process for constructing an answer. By asking for an opinion, the interviewer is unable to determine why the respondent answered *some of the time* and, therefore, is unable to determine whether or not the question was answered as intended. The respondent clearly believes that she answered the question as intended—that is, her reasoning for why she believes the question is not too complicated. However, it is possible that she does not realize that she misunderstood the question. This information can only emerge if the respondent explains the reasoning behind her answer as opposed to her evaluation of the question. Despite the respondent's report that the question was not complicated, there is little knowledge regarding whether or not the question was truly complicated for her.

As Example 3.2 illustrates, in the role of evaluator, respondents can assume the specific role of a question design expert. Interestingly, the interviewer in Example 3.1 asks the same initial probe question as the interviewer in this next example. While the respondent in the previous example spoke only of her own personal opinions, the respondent in this next example speaks more broadly, incorporating how he believes other respondents will process the question. From this understanding he makes suggestions for improving the questionnaire.

EXAMPLE 3.2 RESPONDENT AS EVALUATOR

Interviewer: What did you think about those questions about barriers and obstacles?

Respondent: Well, I think people are not going to like having all of those questions. They will not want to answer them. One question would be fine. One question alone will elicit what the person is actually having trouble with. That would be good. But I do not know that you need three of them.

Again, as in the first example, the interviewer does not collect information regarding the respondent's cognitive processes nor how they interpreted the question within the context of their own life. In Example 3.2, the cognitive interviewer asks a relatively broad evaluative probe; however, the respondent interprets the probe as a request to evaluate the question as it pertains to the research agenda. The respondent, therefore, provides advice regarding the number of questions he believes is actually necessary for the survey questionnaire.

Example 3.3 demonstrates how a respondent can be asked to specifically evaluate the questionnaire as a whole, and how this evaluation can also result in misleading information. The respondent's ultimate assessment—that the questionnaire asks inappropriate questions—stems from the fact that she did not see the skip instructions. As a result, she was mistakenly filtered into questions that were not applicable. Because she did not realize that she missed the instructions, she saw the problem with the questionnaire as asking inappropriate questions rather than the true problem.

EXAMPLE 3.3 RESPONDENT AS EVALUATOR

Interviewer: First, let me get your reaction to it [the questionnaire] at least as far as how easy or how hard it was to fill out. Was it easy to follow or did you really have to work at it?

Respondent: No, it was easy. The only problem I had was where I got to the point where it said "do you smoke" and I put "no" and they continue to ask you questions about smoking. Also, I had my baby vaginally, but they were asking about C-sections. I never had that so that [part of the questionnaire] should be fixed.

This example illustrates why it may be unfruitful, if not misleading, to ask respondents about their opinion of the questionnaire, as well as the difficulty in filling out the instrument. This respondent asserts that the questionnaire was easy to complete, yet she was unable to successfully navigate through the questionnaire. In fact, in this project, 12 out of 20 respondents made skip pattern mistakes. Nine of those

12 reported that the questionnaire was easy to follow when asked in a manner similar to Example 3.3.

Finally, Example 3.4 illustrates how respondents can be asked to examine their own thought processes in order to evaluate a question. In this particular example, the respondent reported that she takes Depo-Provera shots, a synthetic hormone progestin, every 4 months to prevent pregnancy. In a subsequent question, however, she reports that she and her husband are not practicing birth control. The interviewer identifies this contradiction and realizes there must be a problem with one of the questions. In order to determine the specific problem, she asks the respondent to explain why she gave contradicting reports and why the question led her to misreport.

EXAMPLE 3.4 RESPONDENT AS EVALUATOR

Interviewer: Can I just ask you ... on the previous question, "Are you or your husband or partner doing anything now to keep from getting pregnant?" Did you say yes or no to that?

Respondent: I said ... umm ... no.

Interviewer: You are not doing anything to keep from getting pregnant?

Respondent: Yes.

Interviewer: Okay. *And you are having shots of Depo-Provera every 4 months?*

Respondent: Yes.

Interviewer: And why do you take these shots?

Respondent: Huh?

Interviewer: What are the shots for?

Respondent: They say for not getting pregnant.

Interviewer: It is for not getting pregnant. I am just confused. Then, why did you say no? If you are taking Depo-Provera then are you not doing something not to get pregnant?

Respondent: Yes.

Interviewer: But there is something about the question and you said no.

Respondent: It is supposed to be yes.

Interviewer: See, the thing is, we are trying to find out what is wrong with the questions. And there is something wrong with this question because you are taking Depo-Provera not to get pregnant, but up here something about the question made you say, "Well, I am not doing that." So this confused you. Is it because of all those words in the middle? Or would it be better for you if it just said "are you doing something to keep from getting pregnant?" Is it the way they asked the question? Or do you know?

Respondent: [Reads the question again, out loud.] The answer is supposed to be yes instead of no.

Interviewer: Well the people who are doing the research would hope that you would have marked yes. But, see, there is something wrong with the question if you did not. It confused you. So I am wondering how can we make the question better? What should they have said that you would have said yes?

Respondent: [Reads the question again, out loud.] Okay. I was supposed to mark yes because it said shots. It was supposed to be shots. I made a mistake.

Interviewer: Okay, thanks.

Rather than asking the respondent to explain her reasoning for each answer, the interviewer essentially asks the respondent to evaluate her own thinking, providing a rationale for her contradiction. As the example indicates, this line of questioning was not productive and, as a result, the interviewer was unable to determine why the respondent provided contradictory answers. Frustrated because the respondent was unable to provide the needed information, the interviewer eventually drops the issue altogether.

In contrast to placing the respondent in the role of evaluator, respondents can be placed in the role of story teller. Specifically, they are asked to generate a narrative that depicts "why they answered the question in the way that they did." In telling the story, respondents relay what they believed the question was asking as well as describe the various factors that they considered in forming their answer. This narrative—in its complete form—details the particular context of the respondent's life and the various experiences they considered in order to arrive at their answer. The probes used to elicit a "story-teller mode" typically ask the respondent to speak from their own experience, describing and not evaluating their thoughts.

The next example is an interview excerpt illustrating a respondent as a story teller. In order to place the respondent in the position of the story teller, the interviewer asks specific questions about his experience associated with the topic of the survey question. This prompts him to more fully describe the circumstances for his answer.

EXAMPLE 3.5 RESPONDENT AS STORY TELLER

Interviewer: When going about your daily activities, do barriers or obstacles sometimes prevent you from doing the things you want to do?

Respondent: True.

Interviewer: And what situations does that occur? Can you tell me?

Respondent: Because of my hearing problem, when I am in a group and everyone is talking at once, it is impossible. I just have to check out. And therefore, my groups have ... [pause] ... I no longer have groups.

Interviewer: What else are you thinking about? What other situations?

Respondent: Well, I have to ... like if I am going somewhere like this interview ... I have to find out for sure if there is a handicap parking place. I have to find out if there are steps anywhere and sometimes there are. And I just cannot take them.

By describing his experiences in relation to the topic, the reasoning for his answer and the phenomena that the question captures for this respondent are revealed. From his description it is possible to determine that he considered "barriers and obstacles" to be relatively broad—from tangible items such as parking spaces and stairs to more elusive contexts including situations where multiple people are talking. Although broad, these interpretations are considered within the scope of what the question is intended to measure. The next example, which comes from the same cognitive interviewing study, illustrates how this line of inquiry can reveal when interpretations do not fall within the intended construct.

EXAMPLE 3.6 RESPONDENT AS STORY TELLER

Interviewer: When going about your daily activities, do barriers or obstacles sometimes prevent you from doing the things you want to do?

Respondent: Mmmmm. I suppose so. Yes.

Interviewer: Can you tell me what you are thinking of? Why do you say yes to this question?

Respondent: At my job, at work, I feel that my boss does not like me. He does not give me good projects to work on, just little things. Because of that I can never get promoted. He is holding me back. Honestly, my boss is a huge obstacle in my life!

As Examples 3.5 and 3.6 show by asking respondents to explain their answer, researchers can better understand how respondents interpret questions, as well as identify a spectrum of interpretative patterns that would ultimately make up the survey data. In these two examples, respondents' interpretations varied dramatically; one was within scope, while the other was out of scope. Interestingly, when final analysis was completed, findings revealed that those with physical disabilities were more likely to interpret the question as intended. Because respondents interpret questions from the context of their own lives, it often did not occur to non-disabled respondents that this was a question about access for disabled persons.

While specific interview probes can prompt respondents to generate a narrative, it is not always the case that a respondent will provide the desired information. The type of role assumed by the respondent (evaluator or story teller) also depends on the respondent, regardless of the line of probing. In some instances, respondents may be asked experience-based probes, but continue to give his or her opinions about the question as well as speculate about the problems that other respondents may have. It is the responsibility of the interviewer to recognize this type of information and redirect the respondent through another line of probing to achieve experience-based information. This final example shows how an interviewer uses different probing strategies to ultimately collect the type of information that is needed.

EXAMPLE 3.7 RESPONDENT AS STORY TELLER

Interviewer: Do you have difficulty hearing, even if using a hearing aid? "No difficulty, some difficulty, a lot of difficulty, cannot do at all."

Respondent: Uh, no.

Interviewer: *So why no? What are you thinking about?*

Respondent: That question is really straight forward. I do not see how anyone would have difficulty with that.

Interviewer: Ok. *So tell me why you specifically answered no to that question?*

Respondent: Because I do not have any problems hearing.

Interviewer: *Tell me a little bit about your hearing.*

Respondent: I actually have really good hearing. Yesterday my wife was telling her sister on the phone what she got for my birthday—a golf trip. It was a surprise and I was not trying to hear. My hearing is just good—what can I say. [Pause] So, like I said, no one would have a problem understanding that question.

As Example 3.7 illustrates, some respondents may prove to be more "difficult" than others; they are less likely to readily provide the narrative that explains why they answered the question as they did. In these cases, the interviewer must be creative and try multiple strategies. The interviewer in the above example, for instance, redirected the entire line of inquiry from a general probe to one that specifically asked the respondent to describe his hearing. While this line of inquiry did not exactly elicit the story of "why he answered as he did," the interviewer was able to discern that his answer was indeed accurate and that the respondent likely interpreted the question within scope.

As this last example also illustrates, knowing how to probe in order to obtain the best data possible involves many "on-the-spot" decisions on the part of the interviewer. To be sure, conducting a cognitive interview is no trivial task, requiring the interviewer to (during the interview) determine whether respondent stories are complete, quickly decide upon and ask probes that prompt respondents to complete the story, as well as to assess the quality of information that the respondent relays. The next section addresses the role of the interviewer and who should conduct a cognitive interview.

3.4 THE ROLE OF INTERVIEWER

Cognitive interviewing involves an in-depth interaction between the person asking questions and the person answering them. The previous section illustrated the different roles that can be taken on by the respondent (i.e., question design expert vs. expert on their own personal experiences). This section addresses the role of the interviewer and suggests that this role can also be conceptualized in two ways. The first approach is to view the interviewer as a data collector. The other is to see the interviewer as a researcher.

3.4.1 Interviewer as Data Collector

In the think-aloud approach for data collection, the goal of the interview is to collect information regarding cognitive processes and requires that the interviewer remain as absent as possible so as to not contaminate the respondent's thought process. The assumption is that minimization of interviewer interaction minimizes bias and increases data accuracy and validity. At the very least, this approach requires asking minimal questions that do not lead the respondent in one direction or another. However, it can also include interviewers removing themselves from the interview process. For example, they may ask "what are you thinking?" as opposed to "tell me what you are thinking." The interviewer may also decide to not face the respondent since non-verbal cues may unintentionally communicate an idea.

In a verbal probing approach, the interviewer takes on a more interactive role than in the think-aloud approach. However, the verbal probing approach maintains a similar stance in regard to data accuracy and validity. Particularly when the phenomenon being studied is said to be cognition, concerns about interviewer bias are still present. Because the interviewer is used in a more active way, the effort to avoid bias is sometimes accomplished through standardized and structured probes, such that every interviewer asks the same set of questions in the same way to every respondent (Conrad and Blair 1996; 2004; Tucker 1997). The approach is similar to the method of administering a survey question. Every respondent must be exposed to the same stimuli by hearing the same question in the same manner, otherwise, data are inconsistent and the ability to generalize findings is lost. For example, when testing a question that asks respondents if they have coronary heart disease, each respondent would be asked, "what does coronary heart disease mean to you?" For a question that asks, "in the past 3 months, have you been punched, kicked, or slapped?" each respondent would be asked whether or not they could remember as far back as 3 months. In an example provided by Willis (2005), in which respondents are asked about experiencing abdomen pain, each respondent was asked to point to the part of their body that they believed was their abdomen—whether or not they have ever experienced this type of pain. The study found that, indeed, respondents had varying ideas as to where their abdomen was located.

In this view, the interviewer functions merely as a data collector. Of note, this type of interviewer requires no qualitative methodological expertise. By contrast, the researcher—not the interviewer—designs probe questions a priori, according to anticipated problems with the questions and according to the stages of the four-step model of question response. The general goal is standardization of the interview process, including elements from the specific questions asked of the respondent to interviewer behavior during the interview. The interviewer's job is not to look for contradictions and gaps in respondent explications, but to administer the interview protocol, without deviation, as accurately and systematically as possible.

3.4.2 Interviewer as Researcher

As previously discussed, when the target of investigation in a cognitive interviewing study is conceived as respondents' cognitive processes, contamination of that process is of concern. The role of the interviewer, then, is defined as an unobtrusive one whereby the interviewer conducts the interview in a highly standardized manner. There is no ad-libbing. However, when the target of investigation is conceived as a study of meaning and interpretation as it pertains to life experience, the role of the interviewer shifts to that of a qualitative researcher. The interview is understood as a complex interaction, in which the interviewer plays an integral and active role. The interviewer must, in the course of the interview, assess the information that he or she is collecting and examine the emerging information to identify any gaps, contradictions, or incongruencies in the respondent's narrative. The interviewer, then, uses his or her analytic skills to form additional probe questions so that a complete and coherent story is garnered. In short, the interviewer is a researcher.

The transcript below illustrates how the interviewer-as-researcher approach would differ from the researcher-as-data collector approach.

EXAMPLE 3.8 INTERVIEWER-AS-RESEARCHER VERSUS RESEARCHER-AS-DATA-COLLECTOR APPROACH

Interviewer/Analyst: In the past 4 months, have you experienced pain in your abdomen?

Respondent: Yes, I have indeed.

Interviewer/Analyst: Really. Can you tell me what you are thinking about?

Respondent: My wife and I thought I was having a heart attack. My insides were real sore. I could not sleep. Tossing and turning. My wife took me to the emergency room.

Interviewer/Analyst: And you were having a heart attack? Is that what they told you when you got there?

Respondent: No. No, fortunately. It was just acid indigestion. I did not realize it, but the heart attack is on your upper left side. Not here in the middle. They gave me a Pepcid, and in an hour I was fine. Wish I knew that before I got out of bed and got dressed. But it really was very sore.

Interviewer/Analyst: And when was this? Do you remember when this was?

Respondent: It was December 7, our fortieth wedding anniversary. I think it was the pork chops that I had for dinner, that and the cheese grits and chocolate cake! [laughter]. You know, we were celebrating, so we went all out.

Beatty and Willis (2007) discuss the trend in cognitive interviewing to use the interviewer as more than a non-expert professional that merely asks a set of standardized probes. Building on what Willis (1994) called using interviewers as "detectives," they argue that allowing interviewers the freedom to explore issues that arise during interviews can result in the discovery of unanticipated problems in a survey question. Essentially, emergent probing is lauded as a productive way to uncover problems. A disadvantage to this approach as opposed to the interviewer-as-data-collector lies with the cost of interviewer training. Willis (2005) suggests that interviewers should have interpersonal skills and technical abilities, including familiarity with questionnaire problems, training in probing techniques, and exposure to field interviews. Clearly an investment in interviewers is a bigger demand (e.g., recruiting the right people and paying them commensurately) when they are viewed as researchers and not simply data collectors.

When the evaluation of a survey question extends to an exploration of the interpretation of the question and the construct it measures, the role of the interviewer is squarely that of any other qualitative researcher. In addition to exploring issues as they come up in the interview, cognitive interviewers also maintain the goal of identifying emerging patterns and themes. In this approach to cognitive interviewing, the interview itself may be described as "intensive interviewing" (Charmaz 2006) or as a "directed conversation" (Lofland and Lofland 1995). Either way, it is an in-depth exploration of how respondents understand a survey question. In striving to pick up on and explore themes, interviewers guide the discussion with an eye toward answering analytical questions and filling in conceptual gaps.

It is also incumbent upon the interviewer to detect and explore contradictions and incongruencies in the data. For example, the interviewer must decide whether the way a respondent answered a survey question corresponds to the story he or she provided during probing. In addition, the interviewer should notice any contradictions in the respondent's answers to various survey questions. For example, previously in this chapter it was shown how one respondent answered two survey questions in a contradictory manner. In one question she answered that she was taking Depo-Provera shots and in another she answered that she was not practicing birth control. This prompted the interviewer to seek an explanation. Any inconsistency among answers should be detected and explored in a way that explains how the respondent understood the survey question and why he or she chose a particular response option. This is a theme that runs through much qualitative literature. For example, Strauss and Corbin (1990) argue that interviewers must possess an "awareness of the subtleties of meaning of data" (p. 41), and Rubin and Rubin (1995) suggest that interviewers must be able to recognize and explore words that "have rich connotation or symbolic meaning" (p. 21).

In addition to keen observation and analytic skills, cognitive interviewers must also attempt to see the respondent's point of view and be aware of any tendencies toward preconceived notions. For example, in the transcript above (Example 2.8), the interviewer allows the respondent to relay his narrative in terms of what he was thinking in regard to answering the question. The interviewer refrained from asking how he defined "abdomen" and instead guided the respondent through his own experience to determine the type of pain that he included in his answer. In some instances, respondents are not as forthcoming in relaying the narrative. Interviewers, then, must take a more active—and more obtrusive—role in collecting the narrative.

Below, Example 3.9 illustrates how interviewers must sometimes become more active, influencing the interview process in order to obtain the necessary information.

EXAMPLE 3.9 INTERVIEWERS AS ACTIVE PARTICIPANTS IN THE INTERVIEW PROCESS

Interviewer/Analyst: In the past 4 months, have you experienced pain in your abdomen?

Respondent: Maybe. It is possible.

Interviewer/Analyst: Would you say "yes" or "no"?

Respondent: "Maybe" is not an answer?

Interviewer/Analyst: No, it is not.

Respondent: Then I will say "yes."

Interviewer/Analyst: Can you tell me why you said "yes"?

Respondent: Well, I did not really want to say yes.

Interviewer/Analyst: But you ultimately did, why was that?

Respondent: You made me!

Interviewer/Analyst: You could have said no! [laughing]

Respondent: I suppose I said yes because 4 months is a long time and it is pretty common for anyone to have pain at least somewhere around here.

Interviewer/Analyst: So what kinds of pain are you thinking about?

Respondent: Anything really. Is that not what the question is asking? It could be a punch in the side, having the flu, diarrhea, even being real nervous can give you a stomach ache. It could be serious too, like appendicitis.

Interviewer/Analyst: So you do not remember having any of these things in the past 4 months?

Respondent: If I have to think about it …

Interviewer/Analyst: Do you mind?

Respondent: Oh yeah. Last Friday I went out and … you know … had a little too much. You know what I mean? The next day I was worshiping the porcelain. [laughter] You know what I mean? I could not eat anything. My stomach was in bad shape. Never doing that again!

Often referred to as reflexivity, interviewers must be aware of how they influence the interview process and make this known in their study so that others may evaluate the extent to which the researcher's interests or assumptions influenced the inquiry. This differs from the notion that interviewers must avoid biasing the data. Instead, it refers to reflecting on and noting the manner in which their social location may have impacted the interaction, as well as how the questions they asked of respondents frame and shape the data that is ultimately collected. For example, it is helpful to know whether a response came from a direct question or was spontaneously mentioned by the respondent. This allows study users to assess the degree to which topics arose organically from the respondent or were guided by the interviewer. For similar reasons, it is helpful to know whether a respondent's reply came from a generic probe such as "what were you thinking as you answered this question?" or a specific one such as "were you thinking of the previous question when you answered this one?" The goal is not to take interviewers out of the process or otherwise minimize their "footprint" on the interaction, but to clearly demarcate and understand their role within it.

3.5 CONCLUSION

This chapter has illustrated that data collection—who is interviewed and how they are interviewed—provide the underlying framework for a cognitive interviewing study. Decisions about both sample selection and the interview process itself impact data quality, as well as the types of conclusions that can ultimately be made. For the study to serve as a validity check for survey questions, it is necessary for cognitive interviewing data to capture the specific experiences and perceptions that respondents consider when answering the question. This type of data emerges only as respondents relay the story of "why they answered the question as they did." It is the interviewer's responsibility to elicit, as best as possible, this type of information from the respondent. Although occurring simultaneously with data collection, the next chapter will specifically address analysis. As is the case with data collection, the types of analyses that are conducted establish the types of conclusions that can be made by a cognitive interviewing study.

4 Analysis

KRISTEN MILLER and STEPHANIE WILLSON
National Center for Health Statistics

VALERIE CHEPP
Hamline University

J. MICHAEL RYAN
The American University in Cairo

4.1 INTRODUCTION

Cognitive interviewing studies serve multiple functions toward understanding the performance of a survey question. First, the studies identify various difficulties that respondents may experience when attempting to answer a survey question. Identifying these difficulties allows the survey questionnaire to be improved before fielding. In addition, cognitive interviewing studies are a study of construct validity in that they identify the content or experiences that respondents consider and ultimately include in their answer. Finally, cognitive interviewing studies can examine issues of comparability, for example, the accuracy of translations or equivalence across socio-cultural groups. The type of analytic processes employed within a cognitive interviewing study guides the types of conclusions that can be made. This chapter articulates the process of analysis for cognitive interviewing studies and illustrates how the three research goals outlined above can be met. The method of analysis outlined in this chapter is rooted within the principles of qualitative methodology, specifically, within grounded theory methodology.

Researchers have analyzed cognitive interviews in a variety of ways (e.g., Bolton 1991; Chi 1997; Sudman et al. 1996). Other than Willis (2005), however, who describes how question design problems can be determined through cognitive interviewing, there is little discussion regarding the process of analysis in cognitive interview literature (Blair and Brick 2010). That is, there is little explanation as to how cognitive interviews should be examined and studied to produce reputable findings. In keeping with the basic tenets of scientific investigation, the processes laid

Cognitive Interviewing Methodology, First Edition.
Edited by Kristen Miller, Stephanie Willson, Valerie Chepp, and José-Luis Padilla.

out in this chapter are guided by the need for systematic and transparent analyses. A systematic analysis ensures that no one particular case is over-emphasized and that findings do not appear anecdotal. While one particular case may be highly relevant and require particular attention, discounting other cases produces incomplete or misleading findings. The notion of transparency is critical in that it allows readers to understand as well as to cross-examine the ways in which analyses were conducted and how conclusions were reached. Transparency instills the trustworthiness of a study and the reputability and believability of its findings. These tenets carry through data collection and analysis to the final report that must document the analytic process and present evidence to support findings.

4.2 ANALYSIS OF COGNITIVE INTERVIEWS: OVERVIEW

Before detailing individual steps, it is useful to present an overview of the entire analytic process. As is the case for analyses of qualitative data, the general process for analyzing cognitive interviewing data involves synthesis and reduction—beginning with a large amount of textual data and ending with conclusions that are meaningful and serve the ultimate purpose of the study. For example, Miles and Huberman (1994) describe qualitative analysis as an interactive process of "data reduction (extracting its essence), data display (organizing its meaning), and drawing conclusions (explaining the findings)" (Suter 2012, p. 346). For analysis of cognitive interviews, reduction and synthesis can be conceptualized within five incremental steps—conducting interviews, producing summaries, comparing across respondents, comparing across subgroups of respondents, and reaching conclusions. With each incremental step, a data reduction product is created. A description of each of these steps and the resulting reduction product is presented below:

1. Conducting interviews to produce interview text: collecting narratives from respondents that reveal how each respondent made sense of and went about answering a survey question,
2. Synthesizing interview text into summaries to produce detailed summaries: detailing how and why each respondent interpreted the question as well as how they formulated their answers, including events or experiences considered as well as any difficulties answering the question,
3. Comparing summaries across respondents to produce a thematic schema: identifying and mapping common themes that detail phenomena captured and the process of formulating a response,
4. Comparing identified themes across subgroups to produce an advanced schema: identifying ways in which different types of respondents may process questions differently depending on their differing experiences and socio-cultural backgrounds,
5. Making conclusions to produce final study conclusions: determining and explaining the performance of a question as it functions within the context of respondents' various experiences and socio-cultural locations.

Analytic step →	Data reduction product
1. Conducting	Cognitive interview text
2. Summarizing	Detailed summaries
3. Comparing across respondents	Thematic schema
4. Comparing across groups	Advanced schema
5. Concluding	Conclusions

FIGURE 4.1 Products of data reduction for analytic steps

Although these steps are described separately and in a linear fashion, in practice they are iterative; varying levels of analysis typically occur throughout the qualitative research process.

As each step is completed, data are reduced such that meaningful content is systematically extracted to produce a summary that details a question's performance. In detailing a question's performance, it is possible to understand the ways in which a question is interpreted by various groups of respondents, the processes that respondents utilize to formulate a response, as well as any difficulties that respondents might experience when attempting to answer the question. It is the ultimate goal of a cognitive interviewing study to produce this conceptual understanding, and it is through data reduction that this type of understanding is possible. Figure 4.1 illustrates each step in the analytic process and the synthesized product for each step. The downward pyramid represents the reduction of data as the analyst moves from the raw data of individual interviews to the thematic schema generated by the comparisons of interviews.

The different levels of analysis described above represent not only data reduction, but also movement toward larger conceptual themes; analysis is the simultaneous process of data reduction and knowledge building. Figure 4.1 along with Figure 4.2 illustrates the dual nature of this process. The pyramid in Figure 4.1 points downward because it represents the reduction of data; the pyramid in Figure 4.2 points upward because, as the analyst progresses through each step, explanation regarding a question's performance grows.

While the two processes of data reduction and knowledge production may be heuristically separated, in reality the processes occur simultaneously. In reducing the cognitive interviewing data, the analyst produces a more comprehensive understanding of a question's performance; as analysis is performed, understanding of the

Analytic step →	Tiers of knowledge production
1. Conducting	Individual respondent's explanation of answers
2. Summarizing	Record of respondent difficulties identification of potential themes
3. Comparing across respondents	Identification of "what the question captures"
4. Comparing across groups	Response process differences across groups
5. Concluding	Explanation of question performance

FIGURE 4.2 Tiers of theory building for analytic steps

question-response process becomes more complex and complete. In the beginning, it is only possible to understand how each individual respondent makes sense of and answers the survey question. By the end, individual interpretations are understood as well as how those interpretations relate across groups and within the overall context of the question's performance.

It is important to recognize that, with each step of analysis, a series of decisions must be made by the analyst. For example, in writing summaries, the analyst must decide what information is important enough to be included and which information they feel can be omitted from the summary. In developing a thematic schema, the analyst determines whether an interpretive pattern has emerged across respondents. These decisions are significant as they directly bear upon the ultimate findings of the cognitive interviewing study. For this reason, it is important to consider the ways in which such decision making can be made transparent so that results can be replicated or, at least, theoretically understood. An audit trail, consisting of the analytic products generated from each level of analysis, reveals the types of decisions made by the analyst in order to conduct the study.

4.3 ANALYTIC STEPS FOR COGNITIVE INTERVIEWS

Now that the broader processes of data reduction and knowledge building have been outlined, a more detailed description of each analytic step will be presented. Significantly, the analytic product of each step forms the basis for the next step. Because each progressive step is dependent on the previous layer of analysis, it is all the more essential that the analyst's work be apparent. The data reduction products generated with each step of analysis, specifically, the interview text, detailed summaries and thematic schema, provide this transparency.

4.3.1 Step 1: Conducting the Interview

As Figures 4.1 and 4.2 illustrate, the first analytic step is the interview itself. It may seem odd to conceptualize data collection as an analytic component. However, in the interview, the interviewer must identify relevant pieces of information, asking questions to resolve contradictions and gaps in the respondent's story. The interviewer must determine whether or not the respondent answered the question as it is intended, as well as to collect a complete account of the response process. Because the interview is open-ended and semi-structured, it is necessary for the interviewer to identify if, and when, the respondent has veered from relevant topics. To the extent that the interviewer actively listens, sifting out irrelevant or inconsequential information as well as intentionally redirecting the respondent to pursue the specified line of inquiry, the interview is the first step in analysis.

The below excerpt of an interview illustrates the way an interviewer operates as an analyst, redirecting the respondent when she veers off point as well as asking specific questions to uncover the way in which the question performs. The excerpt is taken from an interview conducted for a project to examine a set of children's disability questions. The respondent is answering questions about her 5-year-old son, Jonas.[1]

EXAMPLE 4.1 ANALYSIS WITHIN AN INTERVIEW

Interviewer: Let us get started. The first question is, "Does Jonas wear a hearing aid?"

Respondent: No, he does not.

Interviewer: "Does Jonas have difficulty hearing?" Would you say no difficulty, some difficulty, a lot of difficulty, or cannot do at all?

Respondent: [She pauses to think] I guess, [she laughs] "some difficulty." It really depends on what you mean by that question. His hearing was tested at school. The school has a series of health-related experts that come in and do all kinds of tests. Last week they had blood pressure tests—on preschoolers! Of course his blood pressure was fine.

Interviewer: So you said he had a hearing test. What did you find out from that test?

Respondent: Yes. There was no problem with his hearing. His hearing is fine.

Interviewer: That is interesting. I am wondering why you answered "some difficulty?"

Respondent: [Silence]

[1] All names used in examples are fictional.

Interviewer: Right after you answered, you said that it depends on what was meant by that question. Do you remember what you were thinking when you said that?

Respondent: Right. You could take the question two ways: about actual hearing, but also about listening and focusing. He has no difficulty hearing. It is just that when you are talking, he is not always listening. That is why I said some difficulty. Because hearing could also mean listening.

Interviewer: Do you have an example of his listening problem?

Respondent: It is the normal stuff. Normal for 5-year-old boys, especially. If you want them to pick up their rooms, you have to ask at least 5 times. I do not know if you have kids, but any time ... put on your pajamas, get dressed, brush your teeth. ... When I say some difficulty, I do not mean it is a problem that he needs to go to a doctor or a psychologist or whatever. It is just a problem for me personally as his mother. I guess I should have said "no difficulty" because now I am thinking that you were only looking for kids with a real focus problem—one that is hyper or attention deficit disorder (ADD) or something like that. And Jonas is not that.

As illustrated in this example, several times the interviewer either redirected or asked out-right questions to uncover important aspects of the question-response process. In conducting an interview in this manner, the interviewer was able to determine the interpretive basis of this question ("listening" as opposed to "hearing"), which reveals the construct captured by this question for this one particular respondent. In addition, the interviewer is able to reveal another interpretive pattern: assessing normal behavior as opposed to health-related problems. Both of these interpretive themes revealed in the interview provide a piece of the overall picture regarding the actual phenomena that the resulting variable would measure.

The respondent–interviewer interaction generated from the interview process is considered raw data. Thus, interviews should be documented by audio or video recordings. Transcriptions may follow, but at the very least there should be a saved recording of the interview. Maintaining the actual interview is critical for two reasons. First, it allows the investigator to go back to the original data source in order to verify potential inconsistencies or to follow up on emergent themes that may not have been identified in interview summaries. Second, interview recordings are part of the audit trail that allows all findings to be traced back to the original data source. Without maintaining the original data, studies lack transparency and sacrifice credibility. Along with recordings, it is also important for the interviewer to take notes during the interview. This ensures that, in case the recording equipment fails, the interview is not entirely lost. Interview notes also assist the interviewer/analyst in producing summaries, the second step of analysis.

4.3.2 Step Two: Producing Interview Summaries

While recordings of interviews consist of raw data, summaries represent synthesized versions of those interviews. Summaries should include not only how respondents answered the survey question being evaluated but also what types of experiences or perceptions were considered and any types of difficulties the respondent may have experienced when attempting to answer the question. Types of difficulties could include needing the question to be repeated or asking for clarification. Interviewers should also note how respondents answered follow-up questions posed by the interviewer. In addition, interviewers should document points of respondent confusion about a survey question, contradictions between their answers to the various survey questions and answers to follow-up questions, and inconsistencies in respondents' narratives explaining how and why they answered a question a particular way. Detailed summaries essentially document the initial, albeit rudimentary, analysis that occurred during the interview and allow the interviewer/analyst to reflect more systematically upon what occurred during the interview. The example below illustrates how summary notes could be written from the interview excerpt presented above.

EXAMPLE 4.2 SUMMARY NOTES OF EXAMPLE 4.1 INTERVIEW

Does [name] have difficulty hearing? Would you say no difficulty, some difficulty, a lot of difficulty, or cannot do at all?

After a long pause, she said, "I guess some difficulty" and laughed. Then she said, "It really depends on what you mean by that question." Immediately, without any follow-up questions, she started talking about her son getting a hearing test at school. When I asked her about the results, she said that his hearing was fine. I then asked her why she answered some difficulty if he had no hearing problem. She saw the question as having two possible interpretations: (1) hearing and (2) listening and focusing. She said, "You could take the question two ways: about actual hearing, but also about listening and focusing." NOTE: THERE ARE TWO POTENTIAL INTERPRETATIONS IDENTIFIED IN THIS INTERVIEW. She based her answer of "some difficulty" only on listening and focusing, not hearing. She said "some difficulty" because she gets frustrated with her son for not doing as he is told. Her response was not based on a developmental or medical problem—just her level of frustration. She said that he is a typical child. In the end, after discussing the question, she said that she probably should have said "no difficulty" because she is now seeing the question as asking about a real health/developmental problem. NOTE: HERE IS ANOTHER POTENTIAL THEME: RESPONDENTS MAY ANSWER THINKING ABOUT NORMAL PROBLEMS OR THEY COULD ANSWER THINKING ABOUT MORE SERIOUS PROBLEMS THAT WOULD REQUIRE SOME TYPE OF PROFESSIONAL INTERVENTION.

As illustrated in the above excerpt, the summary interweaves evidence from the interview, including quotes, with analytic findings and potential directions for analysis. From this vantage point (i.e., within a single cognitive interview) basic response problems, such as recall trouble or misinterpretation, can be identified—errors that can be linked to question design problems and, presumably, solutions to these problems. A limitation to halting analysis at this point can, however, produce an incomplete, even erroneous, picture of question performance. As will be seen below, additional analyses comparing this interview to other interviews can reveal a range of interpretations and a variety of phenomena captured by the question.

4.3.3 Step Three: Developing a Thematic Schema

Once summaries for all interviews are completed, the analyst engages in the third step by identifying common themes across respondent narratives. In doing so, the analyst begins by reviewing all interview summaries for a particular question, identifying common themes, that is, interpretive patterns across interviews. For example, in the summary above, an analyst would likely see that most (if not all) of the interviews contained a general theme regarding the concept on which respondents based their answer. This respondent's answer is based on her child's ability to "listen." The next respondent, however, may have based her answer on her child's ability to "hear." From this comparative analysis of even two cases, it is possible to see that respondents can base their answers on at least two interpretations of the question. Comparing the two interviews with the remaining interviews reveals whether or not other interpretations of the question exist. Themes must emerge directly from interviews. Often themes are identified (at least partially) in the interviews or when the analyst is summarizing the interview. Thus, these three analytic steps typically occur simultaneously. As described in the previous chapter, interviews should continue until new themes are no longer discovered—until theoretical saturation occurs.

The approach described here is rooted within qualitative methodology, tracing back most notably to Glaser and Strauss (1967) and their discussion of *grounded theory*. Unlike deductive approaches to data analysis that aim to test a pre-existing theory, a grounded theory approach is based on inductive reasoning, building theoretical claims directly from empirical observation. In this way, thematic schemas are inductively developed "from the ground up." This inductive process is necessarily iterative, meaning that analysts continuously move back and forth between raw data text, themes, and emerging conceptual claims. Glaser and Strauss (1967) refer to this technique as *constant comparison.*

In many instances, as the analyst compares across cognitive interviews, he or she will find variation within a similar pattern. For example, respondents considering their child's hearing (as opposed to listening) may answer based on differing conceptualizations of "hearing"; some may answer thinking of hearing in a noisy room, while others may think of hearing in a quiet room. As these types of differences emerge, a more complex picture of a question's performance emerges. Because of the complexity, it is helpful to visualize the schema as tree branches. Figure 4.3 provides an illustration of the thematic schema for the hearing question as described above.

While this diagram presents a simple illustration, many questions can generate a more complex thematic schema. Figure 4.4 presents such a schema. For this example, the schema represents activities considered by respondents when answering a disability question about their child's ability to perform self-care.

As depicted in Figure 4.4, there are four patterns under the common theme "Activities Considered": (1) self-care in general, (2) only feeding, (3) only eating,

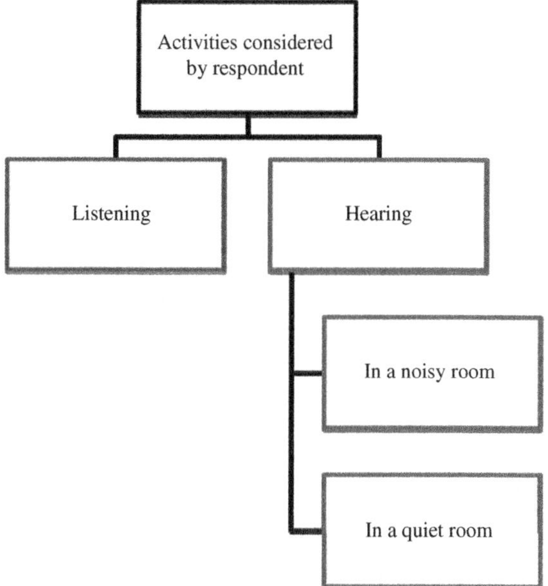

FIGURE 4.3 Visual representation of thematic schema
Does [name] have difficulty hearing? Would you say no difficulty, some difficulty, a lot of difficulty, or cannot do at all?

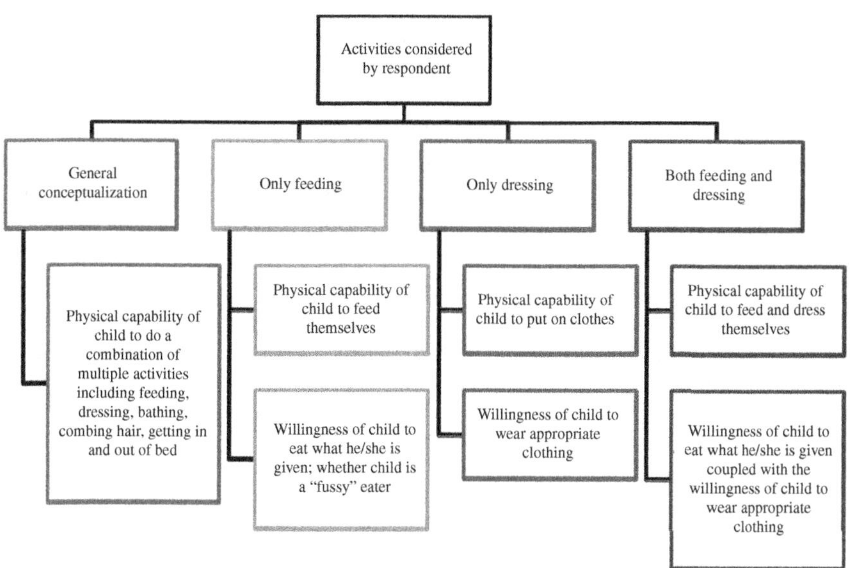

FIGURE 4.4 Visual representation of thematic schema
Compared with children of the same age, does [name] have difficulty with self-care such as feeding or dressing him/herself?

and (4) both feeding and eating together. That is, when answering the question, some respondents considered a generalized conceptualization of health care, while others considered only eating or dressing or both eating and dressing, but no other self-care activity. "Feeding" and "dressing," however, are further divided as various dimensions emerged within these patterns. For feeding, most respondents considered their child's physical ability to eat; however, a couple of respondents considered their child's willingness to eat what they are given. For dressing, most considered their child's physical ability to dress. A couple others, however, considered their child's willingness to wear an appropriate outfit. (For example, one parent explained that his child often insists on wearing his pajamas.) It is the analyst's job to make sense of and explain these patterns. Full explanations generally emerge out of the next analytic step when patterns are compared across groups. In this case, respondents who answered outside the intended scope (i.e., fussy-eating and choosing an outfit) were those with young children who were particularly frustrated with their child; not knowing the intent of the question, their interpretation was framed by their own experience—one of frustration.

The examples provided above illustrate one particular theme found within the interview texts. Typically, however, multiple themes emerge when performing the constant comparative method across interviews. All of the discovered themes can be graphically represented to illustrate the multi-dimensionality of a question's performance. Figure 4.5 below illustrates three patterns that could be discovered through the constant comparative method for the self-care question.

In this case, analysis of interview summaries and interview text reveal three significant themes: "activities considered," "comparisons to same aged children," and "respondent difficulties." Noticeably, the different themes represent key components of the question. The themes also reveal the interpretive dimensions of the question-response process. Examining the thematic schema for a particular question provides important insight into the phenomena captured by the particular question.

The analytic product generated from this step is represented by the thematic schema. Each branch should be grounded and traceable to individual interviews, that is, the original data source. In some cases, particular themes may seem odd or idiosyncratic to specific cases. Explanations for why particular patterns exist, as well as relationships between patterns, are not always evident after the third stage of analysis. In this regard, it is not always possible to determine what, if anything, is problematic with a question. Without fully understanding why a particular problem exists, it is not possible to determine how to "fix" a question. This level of understanding emerges in the next step, when relationships between patterns and groups of respondents are examined.

4.3.4 Step Four: Developing an Advanced Schema

The fourth step of analysis involves systematic examination across groups of respondents to identify whether any particular theme is more apparent among any specific group of respondents. Only by performing cross-group comparisons is it possible to identify the various ways in which respondent experience and social location can

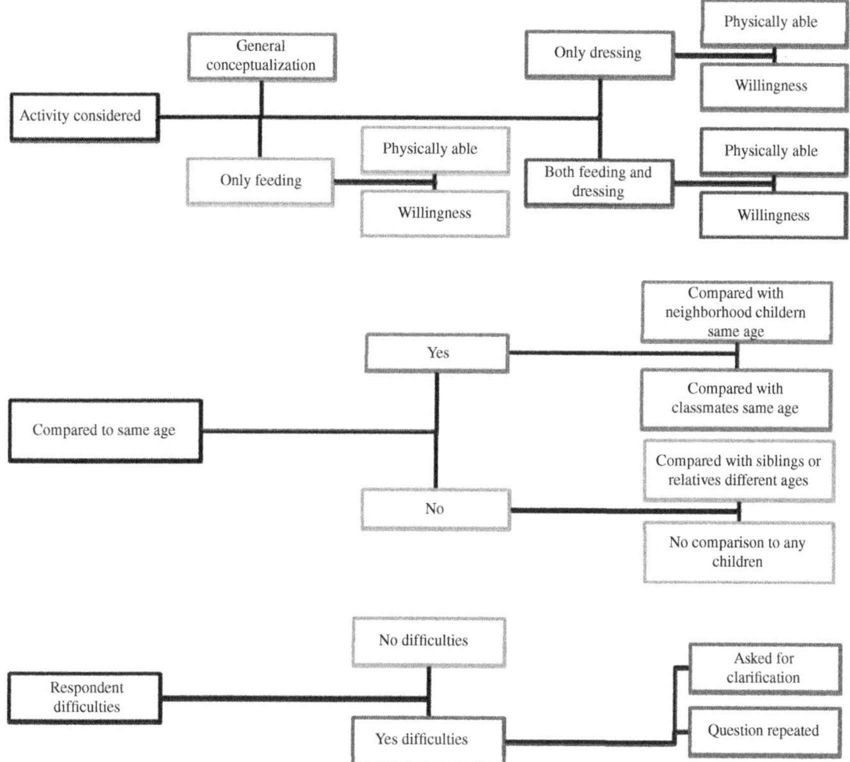

FIGURE 4.5 Entire thematic schema
Question: Compared with children of the same age, does [name] have difficulty with self-care such as feeding or dressing him/herself?

impact question response. It is also the way in which an analyst can better understand and explain why respondents make sense of and process questions as they do. For example, in regard to the children's self-care question, respondents with young children, as opposed to those with older children, were the ones that considered their child's fussiness when answering the question. By comparing themes across groups, an advanced schema is developed, which provides a more complex picture of the ways in which a question performs.

Respondent groups can be conceptualized in a variety of ways. Typically groups are defined by the content of questions that are being studied. For example, in the examination of disability questions, those with disabilities might interpret questions differently than those without disabilities. For smoking questions, those who smoke occasionally might interpret questions differently from those who smoke every day or those who have never smoked. Demographic groups (e.g., gender, education, income, age, and race) are also important to consider because these constructs generate different life experiences for respondents, which can impact interpretation.

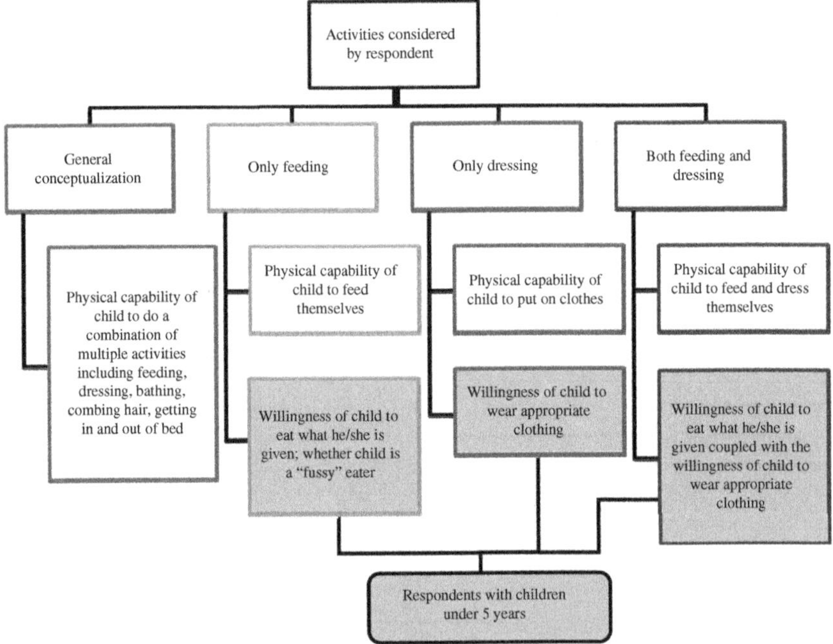

FIGURE 4.6 Advanced schema: comparing across groups
United States/English, Belize/English: Compared with children of the same age, does [name] have difficulty with self-care such as feeding or dressing him/herself?
Oman/Arabic: صعوبة ة .لاأجد صعوبة بـ ن فسكم ذلعندالا سد تحماماوع ندارة داءملابـ سك؟
لاأقدر /لاأستطيعأبداً . صعوبة كبيرة .بـ عضال

If interviews have been conducted in different languages, the analyst should compare across those groups; differences across language groups can indicate a problem with translation. In addition, if interviews are conducted in multiple locations it is important to assess whether regional differences impacted the way in which respondents made sense of questions.

Figure 4.6 illustrates comparisons of the children's self-care question across a variety of different respondent groups. For this question, cognitive interviews were conducted in three countries: United States, Belize, and Oman. The interviews were also conducted in two different languages: English in the United States and Belize, and Arabic in Oman.

For all three countries and for both languages, the activities considered did not vary; patterns for this particular theme were evident in each location and for both languages. These findings suggest that the question is comparable in these specific locations. Importantly, the findings also suggest that the Arabic translation of the question is comparable to the English version of the question. Interestingly, although the constructs did not differ across cultural and lingual groups, it did differ according to the age of respondents' children. Those respondents who based their answer on either their child's reluctance to eat what they were given or on their ability to choose an appropriate outfit were parents of younger children. This pattern was found in all three locations.

4.3.5 Step Five: Making Conclusions

Like each subsequent stage of analysis, the final stage involves synthesizing and summarizing. Many of the conclusions, specifically, the thematic schema and their relationships between groups of respondents, have been identified. However, at this final stage, the analyst must pull findings together to provide factual and relevant conclusions. In essence, making conclusions requires the analyst to bring together the previously developed schema to summarize how the question (or set of questions) performed across study respondents. A description of question performance includes: (1) the ways in which respondents interpreted the question, (2) how they formulated their response, (3) whether there were any difficulties experienced by respondents, and (4) whether a particular group of respondents processed the question differently from others. Ideally, conclusions should also explain *why* the respondent experienced difficulties as well as *why* there are differences occurring across respondent groups. Example 4.3 below illustrates the final step of making conclusions.

EXAMPLE 4.3 COGNITIVE INTERVIEW CONCLUSIONS FOR A QUALITY OF LIFE QUESTION

A cognitive interviewing study was conducted within the context of a health questionnaire to examine the performance of quality-of-life questions. The particular question discussed in this example is: "In general, would you say your quality of life is excellent, very good, good, fair, or poor?"

From the interviews themselves it was clear that not all respondents appeared to put much thought into the question. Others, however, did appear to carefully think through their response. By examining the narratives across interviews (Step 2) it was discovered that respondents with particularly thoughtful responses determined, on their own, three separate components on which to base their response: (1) the particular aspect of their life to evaluate, (2) the time period of their life to evaluate, and (3) a basis for their evaluation. Taken together, these factors established the way in which respondents would answer, suggesting that variation along a particular dimension could alter their response. Some respondents, in fact, requested that their response be changed after discussing their answer. Respondents who appeared to not think through their answer did not consider these components. Unfortunately, it was not possible to characterize the group that did not carefully formulate a response, nor was it possible to *explain why* some respondents formulated thoughtful responses and why some did not.

After understanding the significant components that respondents considered when forming their answers, each of these components were examined. The excerpt below is taken from the final report and illustrates how conclusions were made regarding respondents' interpretation of the question, that is, the construct ultimately captured by the question:

> Construct: "In general, would you say your quality of life is excellent …?
>
> The type of phenomena considered varied across respondents. Approximately two-thirds of respondents based their answer on an evaluation of their health or some combination that included health. In thinking about health, some respondents considered whether or not they had a disease or a chronic condition, while others considered their mental health, their health habits, such as smoking or eating healthy foods, or the types of activities they do. Outside of health, some respondents considered their economic situation, satisfaction of relationships with others, their spiritual life and, for one respondent, his sense of being a "good and moral person."
>
> Although all respondents provided an answer, several respondents could not immediately answer; they were unable to determine what the question was asking and what aspect of their life they should consider. Some respondents were confused because they perceived the questionnaire as being about health but did not see how this question related. For example, when asked the question, one woman stated, "What quality of life? How does that pertain to health? Be more definitive with quality of life." Eventually the respondent answered "excellent" because the "Lord opened her eyes this morning."

For this study, it was also possible to explain how respondents went about answering the question. The below excerpt depicts study conclusions regarding the basis for respondents' answers:

> To formulate a thoughtful answer, respondents were required to devise a basis for performing the evaluation. Typically, respondents made comparisons to another time in their life or to other people. For example, one woman who conceptualized the question as a health question compared her current health state to those of others (i.e., "there are many more people worse off than me"). Another respondent answered "fair," explaining that his financial situation is bleaker now because his construction equipment was stolen and he is unable to work. Prior to this time, he would have answered "good" because he was making "pretty good money." For those respondents who considered multiple constructs at the same time, the question became particularly burdensome because they were required to consider those multiple constructs simultaneously to produce a single answer. Moreover, some respondents did not formulate their evaluation on any particular basis and were unable to provide a sensible rationale for why they chose a particular category as opposed to any other category. For example, one respondent who chose the middle answer "good" stated "… for a 49-year-old male I am doing pretty good. I am getting ready to be 50."

An important rule for an analyst to follow when making conclusions is that findings must derive from the empirical interview data. Expert opinion or self-imposed analyst interpretation—both common pitfalls—should be avoided. For example, a cognitive interviewing analyst might suggest that a question is double-barreled because it (as he or she sees it) has two separate components, when in fact there was no evidence to suggest that cognitive interviewing respondents interpreted the question as such.

Furthermore, the analyst must consider what can and cannot be known about the performance of a question within the context of the specific cognitive interviewing study. In regard to the children's disability questions, for example, it could be concluded that the questions captured the same types of constructs in all three locations. This conclusion holds promise that, on a survey, the questions would capture the same types of phenomena in all three countries. It is also possible, however, that not all thematic patterns were discovered, that "saturation" was not achieved in each of the countries and that, if interviews were conducted in different parts of one or more of the countries, different patterns of interpretation could be found.

Since thematic patterns emerge directly from interviews, the presence of these patterns can be understood as factual—that the patterns do exist. However, the extent to which a particular pattern would exist in a survey sample cannot be known by a cognitive interviewing study. As seen in the self-care question, only some respondents with young children answered outside the intended scope, stating that their child was a fussy eater. The presence of this pattern does indeed exist. However, the extent to which this pattern occurs and the amount of error it would produce if the question was fielded is unknown. Given that this pattern was present in all three countries, and that it "makes sense" that parents of young children would sometimes report these difficulties, a decision could be made to ask this question only of parents with older children. Or, more investigation, such as a mixed method study, could be conducted to determine the extent to which a pattern exists in a larger population. This type of study will be discussed in Chapter 9 of this book.

4.4 THE BENEFITS OF A COMPLETE ANALYSIS

Importantly, conclusions of a cognitive interviewing study should derive from each of the previous analytic steps. Some may argue that a partial analysis of a cognitive interviewing study is preferable to no study at all, that performing half of the analytic steps, provides at least half of the picture. A partial analysis, however, can often create a distorted view of the question-response process and lead to misguided conclusions, especially in regard to what a question captures and how it could be improved. A complete analysis is necessary in order to better capture underlying interpretive patterns and to avoid the possibility of idiosyncratic results. Example 4.4 illustrates the importance of a complete analysis.

EXAMPLE 4.4 FULL ANALYSIS OF A PREGNANCY RISK QUESTIONNAIRE

A cognitive interviewing study was conducted to evaluate a pregnancy risk questionnaire. The questionnaire would be used in a population-based self-administered survey of women with children between the ages of 9 months and 2 years. Topics included health behaviors such as diet, use of vitamins, and smoking. The questionnaire asks respondents to answer questions pertaining to three time periods: (1) the 3 months before the respondent was pregnant, (2) the time period that she was pregnant, and (3) the 3 months following her pregnancy.

Initial analysis of interviews revealed that some respondents seemingly ignored the timeframe specified by each question; women sometimes included events that occurred after the specified time period, a phenomenon known as "telescoping." For example, the women erroneously reported taking vitamins prior to pregnancy when in reality they started taking vitamins only after they discovered that they were pregnant.

Analysis across interviews, however, revealed a more complicated problem than respondents simply overlooking the specified timeframe. Significantly, not all women conceptualized their pregnancies within the three time frames upon which the questionnaire was based. More specifically, women who *did not* plan their pregnancy were least likely to conceptualize their pregnancies in the three time periods and, as a result, were also more susceptible to telescoping errors.

While initial analysis revealed the phenomena of "telescoping," the complete analysis revealed *why* the phenomena occurred. The complete analysis revealed that the primary question design flaw was based on an underlying assumption that all women can and do conceptualize their pregnancies within three specified time frames. Moreover, the complete analysis reveals a potential bias in the resulting survey data: women who are less likely to plan a pregnancy may appear more likely than is truly the case to engage in pre-natal health habits.

The above case study shows how conclusions can vary according to the completeness of an analysis. Moreover, it shows how different conclusions could prompt very different question-design strategies. For example, if analysis consists only of one level of analysis, the researcher may be tempted to conclude that the problem is one of telescoping and add emphasis to the time frame, such as an italic or an underline mark. As this case illustrates, many times individual interpretations are not random or based simply on the grammatical syntax, but are instead interrelated as a function of a common factor or variable. A complete analysis of the Pregnancy Risk questionnaire suggests that the questionnaire could be improved by some structural or sequence changes. Without a full analysis, these underlying patterns are invisible and understanding of what the question measures is incomplete at best and misdirecting at worst.

4.5 CONCLUSION

This chapter has laid out the step-by-step procedures for conducting analysis of cognitive interviews. Although each of the above steps implies data reduction, analysis is simultaneously a process of knowledge building. With each stage of analysis, an analytic product is generated. Each product lays the foundation for the next level of analysis. The products also provide an audit trail which makes the analytic process transparent. By performing a complete analysis, that is, one that incorporates the five incremental stages, it is possible to determine multiple aspects of a question's performance: respondent difficulties, construct validity, and comparability.

5 Assessing Translated Questions via Cognitive Interviewing

ALISÚ SCHOUA-GLUSBERG
Research Support Services

ANA VILLAR
City University London

5.1 INTRODUCTION

This chapter presents arguments showing that cognitive testing provides crucial information regarding how translated questions (as well as other texts used in surveys) are understood, the underlying cognitive processes that respondents undergo to answer them, and how this relates to respondents' interpretation of the same questions in the source language. This information can help improve wording choices in each language used in a study, increase cultural appropriateness of the text, as well as evaluate and improve comparability between translated questions and the original text in ways that cannot easily be achieved by requesting input from bilingual expert reviews.

The first section describes challenges found in multilingual research and different approaches to instrument production in multilingual research. The next section provides some context by describing translation procedures and translation assessment techniques for survey research. The third section reflects on the need for pretesting as part of the translation assessment procedures. The fourth section discusses idiosyncrasies of cognitive testing of survey translations and the fifth section presents examples of problems discovered through the use of cognitive testing of survey translations.

5.2 WHY USE COGNITIVE TESTING IN MULTILINGUAL SURVEY RESEARCH

As it has been described in previous chapters, cognitive testing of questionnaires is an indispensable step in the process to assure collection of high-quality survey data. Yet, most often, in multilingual studies cognitive testing only occurs at best in the

Cognitive Interviewing Methodology, First Edition.
Edited by Kristen Miller, Stephanie Willson, Valerie Chepp, and José-Luis Padilla.

language in which the instrument was designed. The underlying assumption is that, if a question performs as intended in the source language (tapping into the construct of interest, eliciting a codable, accurate, and unbiased response that is supported by the narrative elicited from a respondent) then translators can produce a question version in the target language that also "works" well. However, just as questionnaire designers cannot always anticipate problems with questions that arise when questions are asked of actual respondents, translators and survey translation experts cannot fully anticipate how respondents will interpret and use the translated questions. This section describes the context surrounding multilingual survey research and how the characteristics of this context justify the need for cognitive testing.

5.2.1 Multilingual Research Settings

Multilingual surveys appear in a number of different research settings. Comparative cross-national or cross-cultural research projects, for example, often require asking questions of samples that speak different languages. In research within one country this can also be true if the target population includes speakers of different languages, whether these are countries where native born populations are speakers of varied languages (e.g., Ghana, Switzerland, the Philippines) or where sizable immigrant populations reside (e.g., the United States, France, Germany). The European Social Survey (ESS) is an example of cross-national research where translation is necessary both because it involves countries where different languages are spoken and countries where more than one language is spoken. Each country in the ESS, therefore, produces a version of the questionnaire in each language spoken as first language by at least 5% of the national population: in some countries this 5% represent "official" languages (such as Catalan in Spain), whereas in other countries this 5% of the population speaks a foreign language (such as Hungarian in Slovakia or Russian in Israel) (European Social Survey 2012). In both types of research, comparability of the survey instruments is crucial for the goals of the project.[1]

Comparability can be threatened, among other things, by a number of problems related to how a translated question works. The main reasons why questions might not perform as intended are: (1) problems arising from translation choices, (2) cultural factors influencing interpretation even in "perfectly" translated questions, and (3) lack of construct overlap between the language/culture of the translation and the language(s) for which the questions were designed. We will describe these types of

[1]Survey translation might also take place in two settings where instrument comparability may or may not be a research goal. The first is monolingual research where existing questions that were designed in a different language are borrowed. The second is survey research commissioned by researchers who are not proficient in the language(s) of the target population(s), and thus design the instrument in their own language (as in studies conducted by the United States or European researchers in the developing world). Even when the goal of such studies might not be comparability with another population, cognitive testing of translations would serve the same functions that it does in monolingual research, making sure that the questions as rendered in the language(s) of administration measure what the survey designer intended.

errors in more detail in the section on analyzing cognitive interviews of translated instruments.

Comparability, thus, cannot just be assumed; it needs to be actively pursued through careful research design and assessed by gathering evidence of comparability (or lack thereof) where possible (Van de Vijver and Leung 1997). Cognitive testing of translations is one way to assess how questions will work in the target language(s), what types of errors are present in each language version, and whether questions perform comparably across language versions. Cognitive testing will also provide information about the causes of those problems in question performance that can guide recommendations and point to potential remedies. Thus, failure to conduct cognitive testing in multilingual settings means that the fielded question may not perform in the new language version(s) as intended.

5.2.2 Instrument Production in Multilingual Settings

Depending on the model followed to produce the different language versions, translation may start before or after the source questionnaire is finalized. The most commonly found approach in survey research follows a sequential development, where a source questionnaire is produced and finalized before other language versions are produced (Harkness et al. 2003). Traditionally, this model did not involve translation during source questionnaire production.

More recent applications of this model, however, try to incorporate cross-cultural input earlier in the process, which can involve translation before the source questionnaire is completed. For example, Braun and Harkness (2005) advocated the use of advance translation, a procedure where rough translations are produced before the source questionnaire is finalized as a means of early detection of comparability issues in the questions. Two international survey programs, the International Social Survey Programme (ISSP) and the ESS, have successfully implemented this approach in the past, leading to important changes in the source questionnaire (Dorer 2013). Also in the ESS, draft questions are translated and tested through cognitive interviews and piloted in languages other than the source language before the source questionnaire is finalized (European Social Survey 2012).

The focus of this chapter is on the advantages of using cognitive testing for language versions translated from a *finalized* source questionnaire. This chapter argues that the use of cognitive testing across cultures and languages *while the source questionnaire is still being developed* can help design an instrument that is easier to translate, and avoid complicated situations if similar stimuli cannot be conveyed in a way that grants measurement equivalence unless the source question is modified. This could prevent problematic situations where cognitive testing of translations reveals problems with source questions that are considered finalized and researchers are reluctant to make changes to it based on recommendations arising from cognitive testing or to allow for the necessary adaptations of the target questions.

Other models of multilingual instrument production such as decentering use translation during the source question design stage (Werner and Campbell 1970). In the decentering approach, text is translated back and forth between two (and potentially more) languages until all versions are found to be equivalent; the goal is to avoid questions that are biased toward one of the cultures or languages involved. In this approach, the concepts of a "source" and a "target" question become unnecessary. However, comparison of how questions work with the intended population through pretesting remains necessary in order to overcome the limitations related to expert reviews.

5.3 TRANSLATION AND TRANSLATION ASSESSMENT PROCEDURES

While the focus of this chapter is not to review existing practices for instrument translation, a brief look at survey translation methodology provides a good background for a discussion on testing translated questions. Most current approaches to survey translation advocate the use of some type of team collaboration and pretesting as part of the translation assessment process (e.g., Acquadro et al. 1996; Guillemin et al. 1993; Harkness 2003; Harkness and Schoua-Glusberg 1998; Pan and de la Puente 2005; Van Widenfelt et al. 2005). In particular, the TRAPD model has been discussed and implemented in cross-national survey research programs such as the ISSP, the ESS, and the Survey for Health, Ageing and Retirement in Europe (SHARE).

5.3.1 Team Translation Approaches

As mentioned before, this chapter focuses on research projects where comparability is key and which follow an "ask-the-same-question" model. This model is based on keeping the meaning of translated questions as similar as possible to the source text and maintaining the stimulus constant across languages, while keeping the question format and measurement properties constant (Harkness et al. 2010).

Harkness (2003) describes the five functions that form the TRAPD model: **T**ranslation, **R**eview, **A**djudication, **P**retesting and **D**ocumentation. In this model, two or more individuals translate the instrument into the desired target language. The translators and at least one reviewer then meet to review the original translation(s) and make comments on issues they find or changes they recommend. An adjudicator (who may or may not be at the review meeting) will ultimately decide whether to adopt the changes or recommendations, or make other changes based on reviewer(s)' findings. Then the reviewed translated document is pretested. Throughout the process, decisions made at every step are documented to inform designers and analysts about how the final translation was reached. While other approaches may include all the steps in TRAPD, team or committee approaches have the translation, review, and adjudication steps built into the method.

5.3.2 Translation Assessment Procedures

In ask-the-same-question model, translation is required to render the same questions in all languages and offer the same response options as the original questionnaire (Harkness and Schoua-Glusberg 1998). But how do we test for that "sameness" across languages?

Assessment of translations is embedded in each of the steps of the TRAPD model. In doing their task, for instance, translators worry about a number of factors:

- Is *x* the best way to translate this question (or this term)?
- Will respondents understand this translated question? Is the type of language appropriate for the intended audience?
- Are the nuances of the original question text maintained in the translation?
- Does the wording of the question feel natural?
- Are the response choices likely to be understood and used as in the original question?

Such considerations lead to numerous changes to wording during the translation stage.

At the review meeting, once again, discussion of challenges in the translation process as well as expectations of how questions will perform among respondents happen in a real-time discussion among translators and other experts, and comments will be considered and used to guide review of the translated text.

These *review* steps give translators and researchers answers to translation challenges but only from the point of view of experts (such as linguistic, methodological, substantive, cultural experts). Expert review, however, can yield many different and opposing opinions, leaving the researcher with the difficult task to reconcile them. Thus, translation review is akin to expert review in question design; it will tell us how bilinguals evaluate a translation (Survey Research Centre 2010) but the "true" answers to problems with question wording can only be appropriately explored by talking to members of the target population for the study, that is, by interviewing respondents similar to those who will be administered the translated questions. *Monolingual feedback* (Harkness and Schoua-Glusberg 1998) or monolingual review can be useful to know whether a translation sounds idiomatic and conveys the intended meaning in the target language, as evaluated through a population similar to those in the study. This makes pretesting an essential step in adequate survey translation approaches. Yet, by virtue of being reviewed by monolinguals, it cannot shed any light as to whether the translation seems, even at face value, to be asking the same as the original questionnaire. Therefore input from both bilinguals and monolinguals are necessary for proper translation assessment.

5.3.3 Pretesting as Part of Translation Assessment

Respondents of a translated questionnaire in many cases cannot speak the source language of the questionnaire (or they speak it but live in a country where a different language version is available); we will refer to these individuals as monolinguals, even though they might speak other language(s). Given that they do not know the source language, monolinguals cannot be asked about the extent to which a translated question matches the intent of a source question. Even if respondents happen to speak the source language, their assessment of the translation adequacy would be a matter of opinion, and would leave us no better (probably worse off) than relying on translation expert review. However, respondents speaking the target language can actually provide richer information by allowing the researcher to peek into their cognitive processes as they answer the translated questions, as they elaborate on their answers thereby letting the researcher understand how they interpret each item.

As when testing is done in a monolingual study, the main and primary goal of pretesting translated questions is to uncover patterns of interpretation and respondents' answering strategies, and to evaluate the adequacy of the survey instruments. When testing translations, however, other factors besides those that are strictly semantic must also be comparable in all language versions. Specifically, pretesting needs to investigate many issues that can arise when translating survey questions, related to how these function pragmatically across all languages. Some of these issues are:

- *Differences in social desirability bias.* Questions could be received as sensitive in one language and not in the other, whether due to linguistic or cultural reasons.
- *Differences in level of diction.* One language version might be cognitively more complex to process than the other. For example, in the 2002 ISSP module on Family and Changing Gender Roles, the word "parent" was translated as "progenitor" in Spain in the statement "One parent can bring up a child as well as two parents together." The word was rendered in Mexican Spanish as "padre," which has a stronger male connotation (and hence could be understood as "father") than the term used in Spain. "Progenitor," however, is a much less frequent word and has a higher register than the word "padre," making the question more difficult to understand.
- *Differences in naturalness of language.* Translated questions, while semantically faithful, might not sound idiomatic and be perceived as stilted (Harkness et al. 2004). This may happen because the translator followed the structure or expressions of the original version too closely, thus making it sound like a translation, and therefore changing how the question as stimulus is perceived.
- *Differences in how answer scales and response options are used.* A specific culture's comfort or discomfort with selecting extreme response categories, with expressing disagreement to an unknown person, or with admitting not knowing the answer to a question can affect how questions are interpreted and answered, which may, in turn, influence data quality.

Whether or not these nuances have an impact on measurement is an empirical question. Research looking at response distributions of different question versions, both in monolingual and multilingual settings, suggests that such factors matter sometimes (e.g., Smith 1995; Villar et al. 2006). Therefore, resources need to be assigned to testing translated questions and uncovering potential problems before fieldwork starts. In other words, pretesting methods need to be used for translated instruments as much as for newly developed questions. The remainder of this section presents two possible pretesting techniques that can be used in testing translations, and the next section will focus on the use of cognitive testing for pretesting translated questions.

5.3.3.1 Focus Groups *Focus groups* (Globe et al. 2002) can also be used to involve members of the target population in the assessment process. There are three main advantages of using focus groups as compared to cognitive interviews: (1) they allow the researcher to talk to more people in a shorter time, (2) they permit the inclusion of a varied socio-demographic group of the target population and this provides an opportunity to see if they can reach consensus as to interpretation of items during the group discussion, and (3) they provide a chance to see what is idiosyncratic and what is cultural in language choices and opinions. Focus groups also allow the researcher to listen to how respondents use language among themselves, which can itself illuminate issues of translation. However, focus groups are limited in the number of questions that can be discussed. More importantly, focus groups do not give the glimpse into a respondent's cognitive process in answering questions that a cognitive interview can offer, given that they are not an adequate forum for eliciting respondent-level narrative about the questions they just answered.

5.3.3.2 Open-Ended Pretests Some pretesting approaches provide an opportunity to involve monolingual members of the target population and expose them to translated instrument items to examine how they work for them. In *open-ended pretests*, for instance, the instrument is administered to a respondent without providing response choices, that is, allowing the respondent to answer the question in his or her own words. This approach is useful in detecting translation comprehension problems, and is a good way to elicit possible response choices particularly when researchers are not sure their own lists of choices are comprehensive. But this approach does not uncover interpretation patterns, and this may hide the fact that a translated question is being interpreted in a way different from that intended, even if respondents can easily answer.

5.4 COGNITIVELY TESTING TRANSLATIONS OF SURVEY QUESTIONS

Ideally, cognitive testing of a translated question is meant to uncover patterns of interpretation of a question or text comparable to those in the original language, and both of these should match the interpretation that the researcher intended. Testing should help establish the translated question's validity, that is, how well the question measures the concept(s) it is intended to measure. Even though most times fewer resources are devoted to pretesting survey translations than questions designed in the language of the researcher(s), researchers sometimes agree to conduct cognitive testing of translated questions even though they have not conducted such testing of the original language version. The rationale behind this approach seems to be that while they feel equipped to judge if a question works as intended in their own language, they want independent assessment of the translation quality.

Even though from a practical standpoint testing a translated version should not be very different from testing the source language version, a few particular aspects must be considered when preparing cognitive testing of translated survey texts. This section presents a review of some of the considerations that must be taken into account.

5.4.1 Cognitive Interviewers

The same qualities are needed in a cognitive interviewer in any language. However, it is important to remember a few additional factors. Cognitive interviewers of a translated version would ideally have experience as qualitative researchers and be native speakers of the target language or at least extremely fluent speakers who have lived immersed in the culture of the target population for some extended period of time: they should have near-native-speaker proficiency and "intuition" so that they are sensitive to subtle nuances that other fluent speakers of the language might not understand. The first level of analysis of a cognitive interview is conducted by the interviewer in the course of the interview itself. Thus, a native speaker without adequate qualitative researcher training, or an experienced researcher without such strong language skills, will likely be unable to probe responses adequately to satisfy the needs of that first level of analysis in which inconsistencies across responses or across a response and the corresponding narrative are best explored (see Chapter 4).

5.4.2 Respondent Selection

Cognitive testing participants should be as similar as possible to the target population. An important element in translation and testing of a translation is knowing who the intended target population for the translated questionnaire is. That is, the population for the original instrument may differ in more than just language from those who will answer the translated version. For example, in the ESS, where the original questionnaire is designed in English, the English version used in Great Britain with a general population will differ in language and potentially in cultural aspects from the version used in French with a general population in France, yet both versions need to be aimed at a general population of varied socio-economic status. In the United States, a Spanish version of a national survey questionnaire will be primarily targeting low-income and low-education immigrants from Latin America, predominantly Mexican. Thus, it needs to be translated taking this into account and utilizing language that can be understood by respondents with those characteristics. Other language groups might have a different profile. For example, in the 2010 US Decennial Census the Spanish version needed to work well with a lower-education immigrant population, while the Russian version was targeted for a more highly educated group.

To the extent possible, participants should be monolingual. Bilinguals are, by definition, different from the monolinguals for whom a translation is intended; given that culture is transmitted through language, bilinguals may exhibit different cultural traits when using different languages. As they live across two worlds, they are exposed to the culture of the source language while the target population of the translation may only be tangentially so, or may find it altogether foreign. This is particularly important in populations where some respondents might have a decent ability in the source language. Their understanding of the target language might be different from that of respondents who do not speak the source language, and they might be more tolerant (and less sensitive) to interferences across languages, such as wrong collocations (order or words in sentences) or prepositions.

Within monolingual testing, researchers know that subgroups in the population—subcultures, one may call them—may tend to select some response choices over others, either because their reality warrants so, or because of their own cultural interpretation of the meaning of response categories. Therefore, when selecting participants for cognitive testing, special care must be put into including members from all the relevant subgroups (see Chapter 3).

5.4.3 Introduction, Protocol, and Implementation

Development of the cognitive testing protocol can start from the same protocol as is used for the original language version. This way the same aspects of question interpretation and processing will be considered in all language versions, and it will be possible to examine whether interpretation patterns for questions and response choices are similar across versions. Additional probes can be used as necessary to investigate specifically the interpretation of items/terms that presented particular difficulties or raised uncertainty during the translation process, or what researchers feel might pose cultural or cognitive problems in the target language.

A number of researchers have pointed to challenges faced when implementing cognitive interviewing protocols designed for the general population with immigrants in the United States (Goerman 2006; Goerman and Caspar 2010; Pan et al. 2010), suggesting that cognitive interviewing procedures need to be culturally adapted. However, the Budapest Initiative successfully implemented the same protocol and procedures across seven countries (Fitzgerald et al. 2009; Miller 2008), as do the ESS Question Design Team routinely in the process of designing the source questionnaire (Widdop et al. 2011). In addition, Goerman (2006) acknowledged that: (1) some of the interviewers for the target language had less experience than those in the source language, which might partially explain why probes did not work with respondents who spoke the target language, and (2) native English speakers with lower levels of education also appear to have difficulties with some of the procedures involved in cognitive interviewing. If lower education is a driver of difficulties in participating in cognitive interviews, strategies may need to be followed that improve task comprehension and willingness to cooperate during the interview for the low educated. It is thus possible that to help put respondents at ease, one may need to spend more time explaining the rationale behind the procedure and what it consists of, or providing more examples of what one wants respondents to do.

5.4.4 Analysis

As mentioned before, whether or not the source questionnaire is open to change will determine how problems found in cognitive testing are handled: cognitive testing of a translation can be used as part of the overall instrument design process or inform only the translation. In studies where the source instrument is being tested simultaneously to other language versions, all versions should be tested against the researchers' intended meaning of each question, and results from all cognitive interviews can inform the source questionnaire. If the source questionnaire has been finalized, then the translated text will also be checked against the source text.

TABLE 5.1

	Problems are identified as related to translation choices made	Problems are identified as related to the source question	Problems are identified as related to culture
Willis 2005	Translation errors	General problems	Culturally specific issues
Budapest Initiative (Miller 2008; Fitzgerald et al. 2009)	- Translation error - Interaction between source question and translation	Source question problem	Cultural issues
European Social Survey 2012	- Resulting from translation error - Resulting from source question design	Poor source question design	Cultural portability

One of the ways that cognitive testing can help to evaluate questions (whether original or translated) is through textual assessment. In the original language, when respondents interpret a question in a way different from the intent of the question designers, this signals the presence of a specification error, where the "translation" of the intended meaning into the question wording did not work well. As a consequence, the source question wording needs a revision. In the translated version, when respondents interpret a question differently from the intended interpretation, this could signal either a problem in the original item or a problem with the translation.

Most research on using cognitive testing in multilingual settings uses these three broad categories to classify the types of problems that are found in analysis, albeit with slightly different names. Table 5.1 shows three papers that assign similar labels to the types of problems that are encountered in cognitive testing of translated questions. This classification informs the type of recommendation that will result from the problem: whether the source wording needs to be modified, the translation reviewed, or the entire concept to be measured reconsidered.

It may not be easy to disentangle in a field test of a questionnaire whether an immigrant population responds differently to a translated question about their general health status because they actually are in poorer health than those responding in the original language, because response categories have been poorly translated, or because of cultural factors that make them stay away from self-reporting very positive health status. It is cognitive testing that allows us to examine each of these possible explanations through the analysis of the narrative elicited in testing.

5.5 PROBLEMS UNCOVERED BY COGNITIVE TESTING OF TRANSLATIONS

The following section presents examples of issues that have been uncovered through cognitive testing of translated instruments. Some issues are strictly related to the language version tested, others are likely to relate to problems with the original language version, and others seem to be related to cultural differences between the target populations. These issues are illustrated with examples from cognitive testing of questions either from the literature, or from the authors' own experiences translating survey instruments into Spanish.

5.5.1 Uncovering Translation Problems

5.5.1.1 Uncovering Translation Mistakes Numerous mistakes in translation can be uncovered in cognitive testing: unidiomatic expressions, additions, omissions, and wrong terms can all be revealed in the process of investigating how respondents interpret and answer the question. Example 5.1 illustrates the case of a Swiss French translation where an addition ("par rapport au revenue") made the response options too similar and thus confusing for respondents (Fitzgerald et al. 2009). The added words "in relation to income ratio" made the second response option closer in meaning to the first response option. The recommendation was to add a note for translators in the source question to remind them to check that the two response options were clearly distinct.

EXAMPLE 5.1

Original Item: Using this card please tell me which of the three statements on this card, about how much working people pay in tax, you agree with most?

1. Higher earners should pay a greater proportion in tax than lower earners.
2. Everyone should pay the same proportion of their earnings in tax.
3. High and low earners should pay exactly the same amount in tax.

Translated Item: Veuillez lire les trois affirmations figurant sur cette carte, qui portent sur les impôts payés par les gens qui travaillent. Avec laquelle de ces affirmations êtes-vous le plus d'accord?

1. Les hauts revenus devraient payer une plus grande proportion d'impôt que les bas revenus.
2. Tout le monde devrait payer la même proportion d'impôt par rapport au revenu.
3. Les hauts revenus et les bas revenus devraient payer exactement le même montant d'impôts.

5.5.1.2 Uncovering Regional Variations in Terms Languages are sometimes spoken in different varieties depending on regional differences. For that reason, it has been recommended that the person who translates a questionnaire has the target language as their native language, and that they live in the same region as those who compose the target population (Comparative Survey Design and Implementation Workgroup 2011). For example, to produce a translation of Austrian German it would be best to hire translators and reviewers who speak that variety of German, rather than, say, Swiss German. Even in projects where the language of the source questionnaire is spoken in more than one country, adaptation to regional differences in how the language is spoken might be necessary. For example, differences are found in the American, British, and Australian questionnaires for ISSP modules.

Other projects, however, seek *harmonization* of language versions across countries (see Harkness et al. 2008 for a review of the challenges involved in this approach). If one language version will be used for all countries speaking that language, cognitive testing of the harmonized instrument in each of those countries will be crucial to uncover problematic terms that are not understood or are understood differently.

Regional variations are present also within countries, and linguistic and lexical differences across regions can be more or less pronounced. When translating instruments into Spanish for Latin American immigrants in the United States, for example, researchers encounter varied national origins; language differences across Spanish from different countries are larger than regional variations within those countries, and yet one version in the United States is expected to be used for all respondents, regardless of the variations they speak.

Cognitive testing of Spanish translations in the United States has shown to sometimes uncover the use of terms that do not mean the same across respondents from different countries (Goerman et al. 2013; Schoua-Glusberg et al. 2008; Sha and Park 2013). Conducting cognitive interviews with individuals who speak the different variations might also be more feasible than finding translation experts from each regional variation. Example 5.2 uses an attitudinal question focused on the subject of smoking to illustrate this. During a cognitive interview (Schoua-Glusberg 2005), female respondents from Guatemala explained that "bares" to them evoked the idea of drinking establishments where men go to drink alone or with prostitutes, and where "decent" women simply do not go. The translated item evoked for these women an environment that was not suggested in the original question, and therefore the translation was modified to include more options so as to soften the possible effect of the inclusion of the problematic term. It was not advisable to exclude the term because for speakers of other regional variations it conveyed the intended meaning precisely. In this case, a perfectly valid translation in semantic and linguistic terms was not appropriate from the pragmatic point of view, and would have most likely elicited a different type of response from female respondents from Guatemala as compared to female respondents from other regions where Spanish is spoken and from respondents in the source language.

EXAMPLE 5.2

Original Item: In bars and cocktail lounges, do you think smoking should be allowed in all areas, some areas, or not at all?

Translated Item: En bares o barras ¿cree que se debería permitir fumar en todas las áreas, en algunas áreas, o no debería permitirse del todo?

Modified Item: En bares, barras, cantinas o tabernas, ¿cree que se debería permitir fumar en todas las áreas, en algunas áreas o no debería permitirse del todo?

5.5.1.3 Uncovering Lack of Familiarity with Vocabulary and Idioms Sometimes, a perfectly translated question that semantically means the same as the source language is not understood as intended by the target population. Example 5.3 illustrates this.

The tested question was a Spanish translation of a question asking respondents to say to what extent they agreed or disagreed with the statement "smoking is physically addictive" (Schoua-Glusberg 2005). After asking the question in the cognitive interview, the interviewer probed respondents to provide more information about their answer choice. In doing so, some respondents revealed that they had answered even though they were not sure what "adicción" meant. Others indicated they knew what that term meant but not when qualified with "física." Yet other respondents thought the phrase "adicción física" meant that the body is somehow addicted to the movement of the arm to bring the cigarette back and forth to the mouth, rather than the physiological reaction as the researcher intended. While Latin Americans with higher education are typically familiar and comfortable with this phrase, the translation choice did not work adequately for the study's sample of a general population of Latino immigrants in the United States.

EXAMPLE 5.3

Original Item: Smoking is physically addictive.

Translated Item: Fumar causa adicción física.

Modified Item: Fumar causa adicción a la nicotina.

5.5.1.4 Question not Understood as Expected by the Experts Example 5.4 illustrates how cognitive testing can uncover interpretation issues in translated instruments. Initially "investigational drug" was translated as "droga investigacional" to keep the stimulus as complex as translators felt it would be for respondents to the English version, and with the understanding that this would be the phraseology that might be utilized in a consent form prior to mass vaccinations for community residents in neighborhood schools in case of a hypothetical outbreak of a serious and contagious disease. That term posed problems for a few of the Mexican respondents who interpreted the term "droga" as referring to illegal or street drugs. The recommendation was made to change "droga" to "medicamento" (medication or drug—without the connotation of illegal drugs). After broader consultation with the researchers, and to make the full phrase clearer in Spanish, the term was changed to "medicamento bajo investigación."

EXAMPLE 5.4

Original Item: If you were asked to *sign* a piece of paper at the school saying that the smallpox vaccine is an *"investigational drug"* that has not yet been completely tested, would you be not at all worried, slightly worried, moderately worried, very worried, or extremely worried?

Translated Item: Si en la escuela le pidieran que *firme* un papel que diga que la vacuna de la viruela es una "*droga investigacional*" que todavía no ha sido completamente probada, ¿no estaría nada preocupado(a), estaría un poco preocupado(a), moderadamente preocupado(a), muy preocupado(a), o extremadamente preocupado(a)?

Modified Item: Si en la escuela le pidieran que *firme* un papel que diga que la vacuna de la viruela es un "*medicamento bajo investigación*" que todavía no ha sido completamente probado, ¿no estaría nada preocupado(a), estaría un poco preocupado(a), moderadamente preocupado(a), muy preocupado(a), o extremadamente preocupado(a)?

5.5.2 Uncovering Problems with the Source Question

A review of published reports from multilingual projects where cognitive testing is carried out suggests that problems with the source question might be more common than translation errors or issues of cultural portability. Just one of many issues that can be discovered through cognitive testing of translations is presented below.

5.5.2.1 Uncovering Cognitively Difficult Questions Questions that are cognitively difficult to process include those where something in the item impairs processing, such as text length, syntactic complexity, or grammatical structure. Questions in all languages should elicit a similar level of cognitive difficulty. However, translation sometimes introduces elements that make the question in the target language more complex than the source question. At other times, items that function well in the general population who speaks the source language may not work well among the population that speaks the target language because the latter have little or no formal education (as is the case with many immigrants from Latin America to the United States).

Example 5.5 illustrates an example where the complexity of the word did not change in translation, but where the question proved very difficult to process for some Spanish language respondents. During the cognitive interviews, several respondents asked for re-reads and visibly made efforts to comprehend the question fully, such as frowning. The recommendation after testing was thus to simplify the source question. In fact, sheer length and packing of too much information in the item seemed to cause the trouble with this question, both in the translated and the source version. The original English item had not been cognitively tested, and it was only through testing the translation that designers became aware of the question's cognitive complexity.

EXAMPLE 5.5

Original Item: How much do you think the people making plans to deal with terrorist attacks in your community know about the concerns you would have and the information you would want in these sorts of situations? Do you think that planners know a great deal about your concerns and information needs, a lot, a moderate amount, a little, or nothing at all?

Translated Item: ¿Qué tanto cree que las personas que están haciendo planes para enfrentar ataques terroristas en su comunidad saben acerca de las preocupaciones que usted tendría y la información que usted querría tener en esos tipos de situaciones?¿Cree usted que las personas a cargo de los planes saben muchísimo, mucho, algo, un poco o nada con respecto a sus preocupaciones y a la necesidad de información que usted tendría?

5.5.3 Uncovering Problems Related to Cultural Differences – Need for Adaptation

5.5.3.1 Uncovering Culturally Unknown Concepts In the previously cited cognitive testing study on smoking, it became apparent that although the wording of one particular question presented no interpretation problems, the question expected respondents to have some background/knowledge about cigarette tax. Specifically, the question expected respondents to know that part of the price of a pack of cigarettes are taxes, how taxes relate to services, and what a tobacco control program is. This was found not to be true for all the immigrants with whom the question was tested. Answering the question was difficult for those whose degree of acculturation/education did not allow them to connect the question with any prior understanding of how taxes work in the new context. If there is no possible translation that could fix this difference in how the question is understood, researchers need to resort to other strategies that in most cases will imply either modifying the source question or adapting the question so that it is processed and understood in similar ways in all languages. In Example 5.6 (Schoua-Glusberg 2005) this might mean adding an explanation so that respondents learn about taxes in the culture. Because the source question was not open to changes and the goal of the researchers was to keep the questions as similar as possible, the researchers chose to break down the long question and set up the hypothetical situation in the first sentence. The question itself then became simpler to process.

EXAMPLE 5.6

Original Item: How much additional tax on a pack of cigarettes would you be willing to support if some or all the money raised was used to support tobacco control programs?

Translated Item: Si los impuestos que se cobran en cada cajetilla de cigarrillos se usaran en parte o totalmente para contribuir a programas de control del tabaquismo, ¿qué aumento en el impuesto apoyaría usted?

Modified Item: Supongamos que los impuestos que se cobran por cada cajetilla de cigarrillos se usaran en parte o totalmente para programas de control del tabaquismo. ¿Qué aumento en los impuestos estaría usted dispuesto(a) a apoyar?

5.5.3.2 Uncovering Relevance of Concepts Some questions travel better than others. Oftentimes adaptation is necessary for a question to be pragmatically appropriate and comparable in a culture other than the one it was designed for. As mentioned before, adaptation can also be necessary in the absence of translation, which points to the need to also conduct cognitive testing in new cultures that speak the language of a source questionnaire.

Questions that include examples most often need to be adapted so that the examples are as relevant to the target culture as the ones mentioned in the source question are to the source culture. Cognitive testing provides a great opportunity to test whether respondents from the target population interpret and process the adapted questions in a similar way to the population for whom the source language version was designed. Willis (2005), for example, reports testing an adaptation in a question about food consumption; they found that using "quesadilla" instead of "lasagna" as an example of food containing cheese worked adequately for the target population.

5.6 CONCLUSION

This chapter presented the goals and aims of cognitively testing translated questions. Just as in cognitive testing of source questions or other survey materials, uncovering interpretation patterns and response processes is crucial to assess the validity of translated items. Furthermore, cognitive testing provides a means for assessing a translation's performance and quality, in semantic terms as well as in terms of how it works in the culture and context for which it is intended. In designing and testing a question, a great deal of effort and attention goes into making sure that the phrasing and terms are interpreted as intended and that the item functions well in eliciting the information it seeks. Yet this same focus and care are often not exhibited in producing a version of the question in a different language (and therefore, for use in a different cultural context). Cognitive testing of every language version that will be fielded is unquestionably best practice. It is then possible to answer whether a question is being interpreted as intended in every language in which data will be collected.

The dangers of not using appropriate pretesting techniques to evaluate translated survey tools are not just the same as the dangers of not pretesting questions in monolingual studies. In addition to uncovering unexpected and undesired patterns of interpretation, cognitive testing of a translated instrument can uncover translation issues that might go unnoticed in other translation assessment processes such as expert reviews. Furthermore, it can be used to test and try out different solutions to translation challenges when the review meeting has not led to an unequivocal choice.

A naive observer may wonder if such care should go into the translation process when the questions will undergo cognitive testing anyway. They may argue that if there are question translation issues, these will be uncovered in the elicitation of interpretation patterns; for instance, if a question has a term that has been translated erroneously, the interpretation of the item will point at that error. This is a true and a valid point. In a world with unlimited resources, both time and money, cognitive testing could begin with a very rough draft of a translation. As interpretation patterns are uncovered that point at issues in the draft translation, revisions could be made to the translation and a new round of testing could follow. This process would likely take a number of cognitive testing rounds until all translation problems were uncovered and resolved. In the real research world, however, a better use of resources calls for having the best possible drafted questions before testing begins, both in the original language and in the translated version. Research is needed to investigate what each assessment step and pretesting technique can contribute toward ensuring comparability of different language versions: what kind of mistakes can each uncover, prevent, and help fix.

6 Conveying Results

VALERIE CHEPP
Hamline University
PAUL SCANLON
National Center for Health Statistics

6.1 INTRODUCTION

Conveying research results is a necessary part of the scientific process because it documents, in a transparent way, how research findings are attained. This includes a detailed discussion of how data are collected and analyzed, and how conclusions are drawn. By documenting research findings and making them available to a wider public, stakeholders are able to assess the quality of the empirical data produced. One way researchers convey results is by making their findings available to the scientific community through reports, journal articles, and other forms of published material. Given that cognitive interviewing practitioners and audiences tend to work in applied settings, cognitive interviewing results are typically written up in a report format for an audience of survey professionals. The aim of cognitive interviewing reports is to present findings on how survey questions perform. By presenting these findings, as well as providing information on methodological issues raised or discovered during a testing project, cognitive interviewing reports also help readers obtain specific information about how a new questionnaire testing project can be designed in order to build on the results of previous projects. Depending on the goal(s) of the cognitive interviewing study (see Chapter 3), reports can focus on different aspects of question performance, including difficulties respondents experienced when answering a question, a question's construct validity, or issues of question comparability (e.g., comparability across socio-cultural contexts or translations).

There are two primary types of survey professionals that use cognitive interviewing reports: survey managers and data users. Survey managers are those who are involved in the survey design process; they represent a survey's sponsoring office, agency, or company. Data users consist of anyone who uses the survey data, including academic researchers and policy makers. Different audiences use reports in different ways. For

Cognitive Interviewing Methodology, First Edition.
Edited by Kristen Miller, Stephanie Willson, Valerie Chepp, and José-Luis Padilla.

example, while survey managers and data users are both interested in a question's construct validity and whether a question captures what it purports, survey managers use reports to learn about problems identified in the question so that they can modify questions. Data users, on the other hand, are more likely to use reports as a tool to interpret the data collected after a survey has been fielded.

To date, little attention has been devoted to the topic of conveying results of cognitive interviewing studies. This void in the literature is particularly apparent when compared to the wealth of material published on how to convey results of other qualitative research. These methodological discussions are extensive and have been taking place for decades across different qualitative fields including in-depth interviewing (Weiss 1994; Rubin and Rubin 2012), ethnography (Van Maanen 1988), autoethnography (Ellis and Bochner 2000), and case studies (Yin 2009), among others. However, cognitive interviewing researchers have begun to recognize this important yet often overlooked dimension of the methodology. For example, Boeije and Willis (2013) outline the Cognitive Interviewing Reporting Framework (CIRF) as one systematic approach to writing cognitive interviewing reports, asserting that consistency in reporting mechanisms, such as following the CIRF model, allows for comparability across various cognitive interviewing approaches.

This chapter builds upon these burgeoning discussions taking place in the field and is organized into two parts. The first half of the chapter presents the basic structure of a cognitive interviewing report. The second half explores two key features that are central to a well-written cognitive interviewing report: transparency and reflexivity.

6.2 CONTENTS OF A COGNITIVE INTERVIEWING REPORT

There is no single way to structure a cognitive interviewing report. However, all cognitive interviewing reports should consist of a basic series of five clearly marked sections: an introduction, a summary of the findings, a description of the methods, a detailed question-by-question review, and an appendix. Each section should be concise yet thorough. This organizational format results in a document that is easy to read and accessible across different audiences. The contents and purpose of each section of a cognitive interviewing report are described below.

6.2.1 Introduction

The purpose of the introduction is to orient readers to the research project. This should include an overview summarizing each of the following sections in the report. The introduction should also describe the purpose of the study and any research question(s) motivating the study. If the cognitive interviewing project is being conducted for the general purpose of evaluating survey questions and learning how respondents interpret survey questions, authors can indicate that the purpose of the research is exploratory rather than targeted toward a specific research question. Consider the following example from a National Center for Health Statistics (NCHS) project that evaluated questions from the 2012 Asthma Supplement of the National Ambulatory Medical Care Survey (NAMCS) (Willson 2012).

EXAMPLE 6.1

Division of Health Care Statistics (DHCS) 2012 NAMCS Asthma Management Supplement Study: Results of cognitive interviews conducted August–September, 2011

Introduction

This report documents findings from an evaluation of the NAMCS 2012 asthma supplement. The purpose of the asthma management supplement is to collect information about physician clinical decision making about asthma management for patients in ambulatory care settings. The Asthma Management Survey is sponsored by a collection of federal agencies collaborating on implementing the National Asthma Education and Prevention Program Guidelines for the Diagnosis and Management of Asthma. The goals of the Asthma Management Supplement are to (1) evaluate physician agreement with core elements of the Guidelines, (2) assess self-reported competency in providing Guideline-compliant care, (3) determine which elements doctors report providing, and (4) assess perceived barriers to providing the core elements of asthma management to patient populations. These data will be used to develop interventions to better educate and equip physicians to fully implement the Guidelines.

This is a study of how respondents (physicians) complete the form, how they understand the questions, and what problems if any they encounter in the process. It is hoped that information from this study can inform changes to question design and provide analysts documentation of the underlying constructs measured by each item.

The next section briefly describes the qualitative methodology of cognitive interviewing, including the procedure for sampling interview respondents, the data collection method, and analysis plan. The third section of the report presents a summary of general findings, followed by a more detailed item-by-item review.

In this case, the introduction explains the unique purpose of the cognitive interviewing study in a concise yet thorough manner. It also provides information about the sponsoring agency and how the data will be used, and concludes by outlining the remaining sections of the report.

6.2.2 Summary of Findings

Reports should include a summary of findings that highlights and synthesizes the main findings of the study; other terms for this section include "executive summary" or "analysis highlights." A reader who wants to know the results of a cognitive interviewing study, but who might not be interested in or have the time to read the question-by-question review, can learn about a study's main conclusions, synthesized across all the questions, in this section of the report. The information presented here is a product of higher-level analysis, resulting in more advanced and thematic findings (see Chapter 4). Findings presented in this section could include a vocabulary term used throughout the instrument that respondents interpreted differently, thereby influencing survey responses, or a pattern of interpretation specific to a particular group of people, based upon a social experience specific to that group. Consider the following example from a cognitive interviewing study of the 2010 Youth Traffic Safety Questionnaire, a survey about the driving and traffic patterns and behaviors of teens and young adults (Willson and Gray 2010).

EXAMPLE 6.2

Cognitive Interview Evaluation of the 2010 Youth Traffic Safety Questionnaire

Analysis Highlights

Questions asking about "community" and "typical week"

Items in the questionnaire that ask respondents to think about either their community or a typical week presented interpretation difficulties for some respondents. However, analysis revealed that this pattern did not occur at random. Rather, difficulties were much more likely to be observed for respondents who lived in more than one community or who had no single (and, therefore, no typical) driving pattern.

In exploring this phenomenon further, we discovered that college students who live away from their parents' home for some portion of the year constituted the group who experienced the most difficulty interpreting "typical week" and "community". Because of enrollment in colleges, universities and other vocational programs, college students are in states of residential flux and, therefore, viewed their communities or residencies as essentially split between their "school" address and their "home" address. The distinction between "temporary" and "permanent" home address is one that institutions of higher learning reinforce as well, as they frequently ask that students provide them with both temporary addresses and permanent addresses, and the student necessarily bring this experience to bear when interpreting the survey questions. Many of the items in this instrument ask respondents to assess the behavior of others in their "community," or they ask them to comment on the behavior during a "typical week." Respondents expressed frequent confusion over these questions, and they wondered if they should base their answer on their "permanent" address, generally where their parents or other legal guardians reside, on their temporary school address either on campus or in off-campus housing where they currently live, or perhaps on even the location where they spend the most time during the year. Not only did the way that respondents conceptualize their "home" and "community" vary across respondents, but it also varied for respondents within their interview narratives. In other words, at times they would base their answers on their current "temporary" school address and then later shift to basing their answers on their "permanent" address. Those who shifted their orientation within the interview often chose the location that best "fit" the question at hand. However, this problem of inconsistent definitions of home and community was not found to be a problem among college students who were living at home while being enrolled. This provides further support for the claim that college students present a unique problem for many of the questions on this instrument.

There were some questions where this issue was particularly pronounced. The problem was obvious with the first two questions since they both directly ask about current living arrangements. Even comparing these two first questions, however, it becomes apparent that respondents provided inconsistent answers; for example, those who responded "with parents" but then later answered that they were living in a dormitory or in an off-campus apartment. As mentioned above, questions that used the phrase "typical week" encountered this problem. Questions (1) and (6) in the section on vehicle use, as well as question (1) in the section on cell phone use all must contend with this issue, and the specific problems that arose are detailed in the question-by-question analysis below. In addition to these, there are a number of questions throughout the instrument that ask respondents to compare the behaviors of other drivers and other potential drinkers to their own, and these questions again required that respondents first pick a community on which to base their comparison. These questions are analyzed further in the question-by-question analysis.

In this example, Willson and Gray (2010) highlight the ways in which questions throughout the survey posed difficulty for particular groups of respondents, namely, those who lived in more than one community and those who did not have a predictable driving pattern. In the context of this survey's questions about driving patterns among youth, Willson and Gray (2010) found that college students living away from their parents' home constituted one such group that had difficulty interpreting terms such as a "typical week" and "community."

6.2.3 Methods

Cognitive interviewing reports must also include a detailed methods section that contains information about the sampling and interviewing procedures, as well as the methods of analysis. A transparent discussion of methodological procedures allows readers to assess the credibility of the findings.

Information about the sample should include a description of recruitment criteria and a justification for how these criteria were determined. For instance, if a cognitive interviewing study aims to test survey questions about people's experience with cigarette smoking, it is important to recruit respondents who smoke cigarettes. This section should also discuss how respondents were recruited. For example, was an advertisement placed in a newspaper? Was there a flyer posted in an office building or on a college campus? Recruitment procedures should match the type of sample being sought after. For instance, if a sample of public health professionals is needed, it might make sense to recruit at public health schools or in public health work environments rather than posting a flyer at the local mall. This section should include any sampling issues or concerns that arose during the project as well as a discussion of remuneration or other incentives offered in exchange for participation. Finally, this section should include a demographic overview of the final sample. This is easily done in a table format. A demographic overview is useful for giving readers a sense of the sample's composition and size. Although sampling decisions in qualitative research projects such as cognitive interviewing are often driven by methods of theoretical sampling rather than demographic sampling (see Chapter 3), a quantitative description of the sample provides readers with some information on how varied or homogeneous the sample is according to common identity markers such as race, gender, and class. Because respondents' social locations inform how they interpret the world, demographic diversity—to the extent that it is possible—is desirable in that a more varied sample can offer more varied interpretations and thus a broader foundation on which to build a theory.

Consider the following two examples from the cognitive interviewing evaluation of National Health and Nutrition Examination Survey (NHANES) audio computer-assisted self-interviewing (A-CASI) reactions to race questions (Ridolfo and Schoua-Glusberg 2009). A-CASI is a technology that allows respondents to listen to pre-recorded survey questions and respond to them on a computer or tablet device. This mode not only ensures standardization, but also provides the respondent with a sense of confidentiality and privacy because they are not responding directly to an interviewer.

EXAMPLE 6.3

Testing of NHANES A-CASI Reactions to Race Questions Results of Interviews Conducted in November 2008–February 2009

Sample

For this project, a purposive sample of respondents was recruited in order to evaluate the English and Spanish versions of the questionnaire. See Appendix A for a copy of the English questionnaire and Appendix B for a copy of the Spanish questionnaire used during testing. The sample consisted of 59 adults aged 18–59. Twenty-nine of these adults received the English version of the questionnaire and 30 received the Spanish version of the questionnaire. This sample reflected a variety of demographics in terms of gender, ethnicity/race, education level, and income (Table 6.1). It should be noted that the respondents who received the Spanish version of the questionnaire were monolingual or had very limited English language skills, with a low degree of acculturation to the United States. Therefore, this group is representative of first-generation Hispanic immigrants. Recruitment for the study was done using newspaper advertisements, flyers, word-of-mouth, or by contacting participants from past Questionnaire Design Research Laboratory (QDRL) projects. Respondents were paid $40 for participating.

TABLE 6.1 Demographic Summary of Respondents in Total and for English- and Spanish-speaking Respondents

	English interviews	Spanish interviews	Total (%)	
	$n = 29$	$n = 30$	$N = 59$	
Age				
Under 35	12	11	23	39%
35 and over	17	19	36	61%
DK	0	0	0	0%
Gender				
Female	16	17	33	56%
Male	13	13	26	44%
DK	0	0	0	0%
Hispanic/non-Hispanic				
Hispanic	2	30	32	54%
Non-Hispanic	27	0	27	46%
Race/ethnicity				
White	9	4	13	22%
Black	16	0	16	27%
Asian	1	0	1	2%
Multi-racial	2	0	2	3%
American Indian/Alaska Native	0	3	3	5%
DK	1	23	24	41%
Education				
Less than HS	3	23	26	44%
HS/GED	6	6	12	20%
Some college	10	0	10	17%
4-year degree or more	10	0	10	17%
DK	1	0	1	2%
Income				
Under 20,000	9	15	24	41%
20,000 +	18	15	33	56%
DK	2	0	2	3%

In addition to a detailed discussion of the sampling procedures and a table of the sample's demographics, a report's methods section should also describe the interviewing procedures. This includes a discussion on how the data were collected, who conducted the interviews, the location of interviews (e.g., in a lab, offsite in the field), the duration of interviews, and how the interviews were documented (e.g., audio recordings, video recordings, transcripts).

EXAMPLE 6.4

Testing of NHANES A-CASI Reactions to Race Questions Results of Interviews Conducted November 2008–February 2009

Interviewing Procedures

Due to the perception that some of these questions could be sensitive, they are intended to be fielded using an A-CASI questionnaire in the NHANES' Mobile Examination Center. In order to replicate the conditions in which respondents will receive the survey in the field, respondents were initially administered the questionnaire using an A-CASI questionnaire. At the beginning of the interview the respondents were given a brief training on how to use the A-CASI program. Respondents were then administered the race section of the questionnaire on the A-CASI, followed by the cognitive interview. In the interviews, retrospective, intensive verbal probing was used to collect response process data. Respondents were orally administered the proposed question again and in some of the interviews respondents were asked to recall the answer they gave when filling out the A-CASI instrument.[1] Following this, interviewers administered concurrent probes, which asked respondents to recall how they came to their answer. Probe questions included such things as: Why did you answer the way you did? Can you tell me more about that? What do you think this question is asking? All interviews were audio taped; the audio tapes and interview summaries were used to conduct the analysis. Once the interviews were completed researchers then printed out the respondents' answers given during the A-CASI. Comparisons were then made between respondents' answers to the A-CASI and the answers they provided during the cognitive interview.

[1] Interviewers had difficulty getting respondents who received the Spanish version of the questionnaire to recall their answers from the A-CASI, therefore this procedure was only used for the first 10 Spanish interviews. For the remaining Spanish interviews, the interviewer used the printout of answers from the A-CASI during the cognitive interview.

The methods section should also include a discussion of how the data were analyzed. For cognitive interviewing reports that convey results of testing projects rooted in an interpretivist approach, this can include a discussion of grounded theory and the constant comparative approach to data analysis. Below is an example from the cognitive interviewing evaluation of questions on second-hand smoke (Massey et al. 2013).

EXAMPLE 6.5

Testing of the Questions on Second-Hand Smoke for the NHANES.

Data Analysis

Data from this evaluation were analyzed using the constant comparative method of analysis, in which analysts continually compared data findings to the original data (Lincoln and Guba 1985; Strauss and Corbin 1990; Creswell 1998). This involved a process of data synthesis and reduction (Strauss and Corbin 1990; Suter 2012). Synthesis and reduction were carried out in five incremental yet iterative steps: analysts conducted interviews, produced summaries, compared across respondents, compared across groups, and reached conclusions. Specifically, once interviews were conducted, analysts synthesized interview data into summaries, detailing how and why each respondent interpreted the question and formulated their answers. Next, analysts compared summaries across respondents, identifying common themes. Once themes were identified, analysts compared themes across subgroups, identifying ways in which different groups of respondents processed questions differently depending on their differing experiences and socio-cultural backgrounds. Finally, analysts made conclusions, determining and explaining how a question performed as it functioned within the context of respondents' various experiences and socio-cultural locations. With each analytic step in this process, data was reduced and systematically extracted in order to produce a theoretical summary detailing a question's performance. As such, these different analytic steps represent both data reduction and a movement toward larger conceptual themes. Analysts used Q-Notes, an analysis software tool developed by NCHS, to facilitate data organization and analysis.

6.2.4 Question-by-Question Review

After the introduction, summary of findings and methods section, cognitive interviewing reports should include a more detailed findings section, structured by a question-by-question review. The discussion of each question in the question-by-question review can be organized into various subsections, which may correspond to the levels of analysis outlined in the pyramid in Chapter 4 (see Figure 4.2: Tiers of theory building for analytic steps), particularly levels 3, 4, and 5. For example, the first subsection could describe frequencies, explain respondents' "do not know" and "refused to answer" responses, and describe any examples of out-right error. The second subsection might describe patterns of interpretation among respondents, identifying what the question captures; this corresponds with the third tier of theory building. Given the data, researchers might also describe any relationships or themes that are identified across respondent groups, following the fourth tier of theory building. Quotes from the interview data should be used to clarify this discussion as well as provide evidence that these patterns exist. The report author can then pull together these two tiers of findings and present the overall conclusions on the performance of the question, including any potential for bias. These conclusions correspond to the fifth tier of theory building. It is important to note that not all question-by-question reviews will report on all of these analytic tiers; the report author must consider exactly what data and analysis are available and possible. Moreover, the analysis performed is related to the research question of the specific project (e.g., to determine if all language versions of the questionnaire are comparable or to determine if certain socio-cultural groups of respondents interpret questions similarly). The research question must be addressed in this discussion. Consider the following example of a question-by-question review from a cross-national cognitive interviewing study that tested questions about children's disability (Massey 2014); this example reports on findings across multiple tiers of analysis.

EXAMPLE 6.6

Analysis of Cross-National Cognitive Interview Testing of Child Disability Questions

7b-1) Compared with children of the same age, does (he/she) have difficulty learning to do new things? Would you say…

(1) No difficulty (2) Some difficulty (3) A lot of difficulty (4) Cannot do at all

This question was asked to 183 respondents with children between the ages of 3 and 17 in all five countries. Ninety respondents answered "no difficulty." Fifty-eight respondents answered "some difficulty." Twenty-eight respondents answered "a lot of difficulty," and seven answered "cannot do at all." Twelve respondents asked for clarification. Specifically, several respondents were not sure about what was meant by "new things."

Respondents interpreted this question in three ways:

- **Learning contexts:** Most respondents considered the question to be about how easily their child is able to learn to do new things or acquire new knowledge. Almost all respondents focused on either school-related or physical (sports and games) activities. For example, a respondent from Belize indicated that his son had "some difficulty" in this area because "when it comes to his school work it is hard for him to learn math" while a respondent in India answered "no difficulty" because her son learned to skate and to play cricket with ease.

Respondents' narratives revealed a slight tendency toward culturally specific patterns in the activities that respondents focused on. Respondents in Montenegro tended to focus on learning in general while respondents in the United States, India, and Belize focused more on school and sports. Respondents in Oman often focused on cooking and technology (such as computers or electronic games).

For respondents who interpreted the question in this way, difficulties included problems that impeded learning such as the child's need for extra time or repetition to master the new information or ability. Some respondents noted that their children's ability to learn depended on the type of activity. Since some things are easy to learn and others hard, these respondents answered "some difficulty."

Respondents whose children had differing levels of ability in physical and mental domains tended to prioritize mental ability. For example, a respondent in Montenegro said, "The boy is mentally fit. There he has no problems, but when it comes to doing something which requires physical activity, he cannot do it." This respondent answered "some difficulty." In contrast, a respondent in India described her son's ability to learn complicated dance routines after seeing them only one time. This respondent answered "a lot of difficulty" because her son has trouble with his studies and prefers to learn things "when they are taught like a song." Another respondent made a distinction between learning and doing. This respondent answered "a lot of difficulty" even though she described her son as "very good at learning." She went on to say, "There is no difficulty in learning things, but he requires maximum support from someone to execute it." This respondent answered based on her child's ability to apply what he had learned rather than on his ability to learn.

- **Interest/Willingness:** Twenty-five respondents answered based on their children's interest in learning but there was inconsistency in how they formulated their answers. A respondent in India who answered "some difficulty" said her son can learn new things "but only if they interest him." Another respondent answered "a lot of difficulty" because her son "does not like to learn." These respondents answered based on their children's desire to learn rather than on their ability to learn. In contrast, a respondent from Montenegro indicated that his son has "no difficulty" learning. He said, "What interest him, he immediately learns. For example, video games."
- **Anxiety about new things:** A small number of respondents (n = 12) focused on their children's reaction to "new things." For example, one respondent in Montenegro who answered "no difficulty" described her daughter as "very curious" and "always interested in new toys." Other respondents described their children's fear and anxiety in the face of novel situations. Another respondent in Montenegro described his daughter's difficulty learning, saying, "She has difficulty because she has a fear of new things." These respondents focused on their children's reactions to unfamiliar things rather than their ability to learn.

In the context of disability: Respondents whose children are disabled often answered based on the context of the disability. Parents of disabled children have different expectations for their children than for non-disabled children, and for this reason, they do not usually compare their children to "children of the same age." For example, one respondent in the United States indicated that her profoundly mentally and physically disabled daughter has "a lot of difficulty" learning. When asked to describe something that her child had learned recently, the respondent paused for a long time before telling how her daughter had recently learned to take a single step with her trainer. This respondent went on to say, "I did not want to say "Cannot do at all." This respondent did not want to think that her daughter was incapable of learning. The few respondents who did compare their disabled children to non-disabled children answered "cannot do at all." For example, a respondent in India said, "She can learn with much repetition. It must be 10 times more than other children."

Frustration/Persistence: Respondents often characterized their children as either easily frustrated or persistent when trying to learn new things. Respondents who felt frustrated with their children's learning, or whose children were easily frustrated, tended to answer "some difficulty" or "a lot of difficulty." Other respondents characterized their children as "persistent" describing how they "do not give up." For example, a respondent in India said his child "tries to do new things and does not give up until he finishes even if it is hard." These respondents tended to answer "no difficulty" even if their children were slow learners.

In this example, Massey (2014) conveyed the results of her cross-national analysis of this particular question. First, she reported on the frequencies, specifically highlighting respondents' requests for clarification of the term "new things." Next, she described three patterns of interpretation among respondents, illustrating how respondents variously interpreted the question to be asking about their child's ability to acquire new knowledge, their child's interest in learning, and their child's reaction to unfamiliar information (rather than their ability to learn new information). Finally, comparing across parents of disabled and non-disabled children, Massey identified different levels of learning expectations, and how these different expectations variously shaped respondents' answers. Throughout the analysis, Massey used quotes from the interview data to clarify the findings and to shore up evidence for the varying interpretation patterns.

6.2.5 Appendix

The appendix is the final section of a cognitive interviewing report and should include a copy of the survey instrument that was tested. This allows a reader to see the exact set of questions that were tested, as well as any interviewer instructions or standardized probes that may have been used. Additional appendices might also be included if appropriate, including a copy of the translated questionnaire if it was tested in a different language, or other survey materials that might have figured into the testing such as the consent script or letter.

6.3 CHARACTERISTICS OF A COGNITIVE INTERVIEWING REPORT

Just as all cognitive interviewing reports should follow a basic structure, there are also a number of characteristics that every cognitive interviewing report should have in order to establish *credibility*. Credibility, or trustworthiness, refers to the validity of qualitative research (Suter 2012). The credibility of research results relies primarily on the trust the audience has in the researcher. In cognitive interviewing reports, like other qualitative methods, credibility is demonstrated by a researcher's *transparency* and *reflexivity*. The writer demonstrates transparency by providing a clear research audit trail that details methods and procedures (Hiles 2008), and reflexivity by being "aware of the multiple influences they have on research processes and on how research processes affect them" (Gilgun 2010, p. 1). These two characteristics of a credible cognitive interviewing report are described in more detail below.

6.3.1 Transparency

As Hiles and Cermak (2008) point out, transparency is perhaps the most important characteristic of qualitative research and writing. Because a researcher's decisions and social context have a great impact on a qualitative method like cognitive interviewing, research audiences must be provided with a transparent audit trail of how the research was conducted from start to finish. The cognitive interviewing report is an opportunity to document this transparency. This includes everything from the decisions behind choosing cognitive interviewing as the correct method of inquiry, to the specifics of the recruiting and data collection processes, to the analytic methods used to interpret the data and come to the findings. A clear and transparent audit trail allows readers to efficiently and accurately cross-examine and judge not only a researcher's credibility, but also the validity of a researcher's findings (Merriam 2002, p. 21).

While transparency is essential to establishing credibility and validity, validity can be strengthened in qualitative research through the use of constant comparison and a consideration of negative cases, techniques that take place during the analytic stage of the research process (see Chapter 4). These processes should be made transparent through the report-writing stage of the scientific process. By documenting these methods, procedures, and decisions in detail and making them available to a broader public, researchers create a transparent audit trail that others in the scientific community can reference and assess.

Conveying results in a transparent way also entails a commitment to creating channels of dissemination that make cognitive interviewing reports easily accessible and widely available. For example, the National Center for Health Statistics (in partnership with the United States Census Bureau, the Bureau of Labor Statistics, the National Science Foundation, the National Agricultural Statistics Service, and the National Cancer Institute) created Q-Bank, a public database of cognitively tested questions from federal surveys. The database links each survey question to its test findings and the original question evaluation report (for more information, see http://wwwn.cdc.gov/qbank). Both survey managers and data users can benefit from such a database. On the one hand, for survey managers, this centralized clearinghouse of reports and tested questions includes pre-tested questions that survey designers can incorporate into their organization's surveys in an attempt to reduce respondent burden and ensure construct validity. On the other hand, the database provides data users with easy access to reports, which they can use as a tool to aid their interpretation of the quantitative survey data. Further, by sharing cognitive interviewing reports with the public, others in the scientific community can review and potentially replicate the study's procedures.

6.3.2 Reflexivity

In qualitative methods, the researcher is the research instrument. Qualitative research, such as cognitive interviewing, does not use microscopes, MRI machines, or any of the other high-tech, computer-based machines that provide instant and seemingly unbiased or "objective" results. Rather, in qualitative social research, a human being is both the researcher and research instrument; data collected by the researcher is necessarily filtered through the researcher's interpretations of the data. Because of this, it is vitally important for the researcher to reflect upon, and then report, how his/her presence may have affected the results of the research. This reflective process is called reflexivity and is a core component of qualitative methodologies rooted in interpretivist traditions. This includes reflection on how researchers determine some research questions to be more salient than others, how they influence the type of data collected, and how they decide to make sense of the data. Qualitative researchers are encouraged to write down these reflections in the text of interview transcripts, summaries, field notes, or analytic memos. These reflections take place during the data collection, analysis, and writing phases of the scientific process, and they constitute a source of data in and of themselves.

Evidence of reflexivity should not only appear in the analyst's research notes but should also be discussed in published communication formats, including cognitive interviewing reports. While this will look different from the ways in which reflexivity is discussed in other qualitative communication formats (e.g., discussions of reflexivity can constitute a whole chapter in book-length ethnographic studies), varying combinations of the following features of reflexivity might be important to discuss in cognitive interviewing reports:

- A discussion of the interviewer's social location (e.g., race, gender, sexuality) and research training or other credentials,
- Any (subjective) research decisions made along the way, for example, decisions on unplanned probes,
- Any instance of leading probes,
- Potential context effects, which occur when previous questions influence responses to subsequent questions in a survey,
- A consideration of how the physical environment might shape the data collected.

To determine which elements are important to discuss in any given cognitive interviewing report, analysts should reflect upon how, if at all, these factors might have shaped research findings. For example, NCHS, in collaboration with the Census Bureau and Internal Revenue Service, examined questions that aimed to measure respondents' trust in federal statistics. Questions asked respondents to say whether they strongly agree, somewhat agree, somewhat disagree, or strongly disagree with statement such as: "Statistics provided by the federal agencies are generally accurate," "People can trust federal statistical agencies to keep information about them confidential," and "Statistics produced by federal agencies, like the Census Bureau and the Bureau of Labor Statistics, do not favor one political party or another." Given that questions centered around respondents' trust in federal statistics, researchers demonstrated reflexivity in their decision to conduct the majority of the 42 interviews outside the context of a federal statistical agency. Willson (2013), the principal investigator on the project, documented this decision in her report.

EXAMPLE 6.7

Cognitive Interview Evaluation of the Federal Statistical System Trust Monitoring Survey, Round 1: Results of interviews conducted in October, 2011

The Census Bureau, NCHS, and IRS participated in cognitive interviewing. We completed a total of 42 interviews. Some interviews (18) took place in the lab at NCHS. However, the Census Bureau and the Internal Revenue Service conducted the remainder of the interviews off-site in an effort to create an environment conducive to respondents expressing distrust of government and government institutions.

In this case, Willson (2013) reflects upon the possible ways that the physical environment of being in a cognitive interviewing lab at NCHS might shape respondents' answers to questions about their trust in federal statistical data. Importantly, she explicitly documents the decision to conduct interviews off-site in the cognitive interviewing report.

In another case, the Centers for Disease Control and Prevention and the United Nations Children's Fund wanted to examine the sexual behavior and decision making of late teenagers and early adults across the world. A series of cognitive interviews were conducted in Manila, Philippines, and Zomba, Malawi to evaluate the proposed questions. In both locations, cognitive interviews were conducted in both the local languages (Tagalog and Chichewa, respectively) and English. Thinking and writing about how something as basic as the interviewer's language can function as a social sign is a necessary component of demonstrating reflexivity. Consider the following excerpt from the Malawi cognitive interviewing report (Harwell 2013).

EXAMPLE 6.8

Violence Against Children Survey (VACS) Cognitive Interview Study, Malawi: Final report

The Impact of Interviewers
The impact of interviewers on data collection in survey research has been well established. These effects are then magnified when sensitive information is being collected (Tourangeau et al. 2000; Mensch and Kandel 1988; Singer, Frankel, and Glassman 1983; Catania, Binson, et al. 1996). Given the topics covered in

the VACS questionnaire, it is not surprising that we saw examples of respondents editing their answers. The most glaring example of this is exhibited by the fact that no Chichewa respondents reported having sex. In addition, no Chichewa respondents reported that anyone had tried to make them have sex but not succeeded (F800), that anyone had ever physically forced them to have sex (F900), or that anyone had pressured them in a nonphysical way to have sex (F1000).

Aside from the generic impact of interviewers, there were other factors specific to this study that could have caused these effects. One such factor is the impact of interviewer gender, which has been known to have an impact on respondents' answers. All Chichewa interviewers for this study were male, but for English interviews the majority of respondents were paired with an interviewer of the same gender. This could have resulted in female respondents feeling uncomfortable discussing sensitive topics with a male. Another potential factor is the fact that all of the English interviews were conducted by American interviewers. Given that the topic of sexual activities is a particularly sensitive topic in Malawian culture, respondents may have spoken more freely with English interviewers due to the fact that they are an "outsider" to Malawian culture.

Here, Harwell (2013) dedicates a subsection of the report to the impact that interviewers might have had on the research process. Specifically, he reflects on the ways in which interviews' nationality or perceived "outsiderness" versus "insiderness" might have shaped how respondents answered culturally sensitive questions. He demonstrates reflexivity again when discussing the ways that gender might have also shaped the interview data.

Contrary to what one might think, integrating discussions of reflexivity in cognitive interviewing reports does not "expose" the weaknesses of the research process or undermine research findings. Rather, qualitative research traditions grounded in interpretivist epistemologies recognize the multitude of ways in which a researcher invariably shapes the research process. Being reflexive about this reality actually strengthens a research report in that it aids in the credibility of the researcher and the transparency of the research process.

6.4 CONCLUSION

Conveying results is a vital aspect of the scientific process in that it allows various audiences to more fully understand the findings produced in cognitive interviewing studies. Cognitive interviewing reports should follow a basic structure so that readers can clearly understand how conclusions were drawn, based on the evidence provided. Further, reports should be transparent and reflexive, thereby lending to the credibility of the qualitative research findings.

7 Case Study: Evaluation of a Sexual Identity Question

KRISTEN MILLER
National Center for Health Statistics

J. MICHAEL RYAN
The American University in Cairo

7.1 INTRODUCTION

This chapter provides a case study example[1] of the methodological approach outlined in this book. The chapter also illustrates how findings can be written to demonstrate study credibility and trustworthiness. Moreover, this case study illustrates how cognitive interviewing studies are able to explain and resolve question design problems that impede survey data quality. It also illustrates how cognitive interviewing studies can examine construct validity by determining the basis of respondents' answers as well as identifying construct differences across groups. Significantly, this study builds upon existing cognitive interviewing studies, illustrating the benefits of publicly accessible written reports or papers that document question evaluation findings. With accumulated knowledge regarding the question design of specified constructs, as this case study reveals, cognitive interviewing studies provide necessary information for the development of survey questions that can ultimately capture intended constructs.

The goal of this particular project was to develop and evaluate a sexual identity question for the National Health Interview Survey (NHIS).[2] Development and then evaluation of the question was based on findings from cognitive interviewing studies conducted by the Questionnaire Design Research Laboratory (QDRL) at the National

[1]This example is from Miller and Ryan (2011).
[2]The National Health Interview Survey (NHIS) has monitored the health of the nation since 1957. NHIS data on a broad range of health topics are collected through personal household interviews. Survey results have been instrumental in providing data to track health status, health care access, and progress toward achieving national health objectives.

Cognitive Interviewing Methodology, First Edition.
Edited by Kristen Miller, Stephanie Willson, Valerie Chepp, and José-Luis Padilla.

Center for Health Statistics. To determine the most appropriate sexual identity question for the NHIS, seven previous testing projects were conducted. Including this study, a total of 386 in-depth cognitive interviews were collected. (For final reports of previous projects, see Q-Bank at http://wwwn.cdc.gov/QBANK/Home.aspx). In addition, data from the 2002 and 2006 National Survey of Family Growth (NSFG)[3] were examined to further investigate findings from past cognitive interviewing studies.

This chapter presents the cognitive interviewing study conducted to evaluate the final version of the NHIS sexual identity question. The chapter begins, however, by defining the construct to be measured and then outlines known question design problems with existing sexual identity measures. It then presents a revised version of the question and the rationale for the new design. The chapter, then, presents the case study conducted to perform evaluation of the newly revised question. In this section of the chapter, the methods used for the study are presented, including the five-tier approach to analyzing cognitive interviewing data. Finally, a detailed discussion of the findings is presented.

7.2 BACKGROUND

7.2.1 Intended Construct for the National Health Interview Survey Sexual Identity Question

Before designing a survey question, it is necessary to identify the specific construct that the survey question should be capturing. For this project, the intended construct is sexual identity, which must be differentiated from other terms used to characterize the sexuality of populations. The term "sexual orientation" is commonly used in today's lexicon as a catch-all term that does not specifically refer to a construct. The term has come to refer to a combination of characteristics that include a person's history of sexual behavior, how they conceptualize and summarize their attractions toward opposite and same-gender people, and how they have come to understand and label their own selves. These three concepts—attraction, behavior, and identity—although inter-related, pertain to different aspects of sexuality and are typically asked as separate questions in survey questionnaires. In addition, the three differing constructs may be of varying relevance to a particular research study. For example, a study intending to examine the spread of sexually transmitted diseases would likely be more interested in respondents' sexual histories as opposed to the label that individuals use to describe themselves.

There is great interest in understanding health disparities across multiple population dimensions. For this purpose, the construct of sexual identity is the most appropriate aspect of the general concept of sexual orientation. Sexual identity is best conceptualized as a concept of self that is formed within a social context and

[3]The National Survey of Family Growth (NSFG) gathers information on family life, marriage and divorce, pregnancy, infertility, use of contraception, and men's and women's health. The survey results are used by the U.S. Department of Health and Human Services and others to plan health services and health education programs, and to do statistical studies of families, fertility, and health.

defines an individual's relationship to other individuals, groups, and sociopolitical institutions within that context (Rust 1993). Furthermore, identities are instrumental in organizing peoples' lives and their everyday interactions, which hold important implications for individuals' behaviors and others' actions toward them (Cast 2003). In the context of health, sexual identity is informative in understanding respondents' access to health care and, subsequently, the quality of care they are provided. It is also informative in understanding risk factors such as diet, exercise, stress, and smoking patterns, as these factors are closely linked to community as well as self-conception. It is important to note that although individuals may conceptualize their identity within a framework of who they have sex with or who they are attracted to, behavior and attraction in and of themselves do not constitute identity. It is the meaning that the individuals assign those behaviors and experiences that define how they ultimately conceptualize their identity (Plummer 1981, 1995).

7.2.2 Review of Data Quality Problems

Measuring sexual identity on a survey questionnaire presents unique challenges. Sexual identity, as is true of other forms of identity, is a complex concept that is rooted in social and political contexts and can change over the course of an individual's life. Consequently, individuals' sexual identities do not necessarily conform to discrete, objective, and uniformly defined categories. In addition, as past evaluation of sexual identity questions using cognitive interviewing methods revealed, the construct itself can differ substantially across various sexuality subgroups (Ridolfo et al. 2012). While the concept of "sexual identity" holds a particularly distinct and salient meaning for those identifying as lesbian, gay, bisexual, or transgender, many non-minority respondents (i.e., heterosexuals) do not hold salient sexual identities. Instead, these respondents often disidentify from a gay identity, possessing what is referred to as a "not-me" identity (McCall 2003). Rather than identifying as heterosexual, these respondents typically identify as "not gay" or "normal."

This lack of construct comparability may generate relatively disparate data across sexual minority and non-minority groups, though more significantly it generates different types of response patterns. Misclassification and missing data errors can occur, however for different reasons. Non-minority respondents who do not identify with a particular sexual identity are not always familiar with the response categories, specifically, the terms "heterosexual," "homosexual," and "bisexual." For example, previous cognitive interviewing studies found that respondents can confuse the words "homosexual" and "heterosexual," believing that "heterosexual" is the equivalent of being gay and that "homosexual" is the equivalent of being straight. In addition, some cognitive interviewing respondents, not knowing the terminology, surmise that the term "bisexual" means "heterosexual," concluding that "bi" means two: one man and one woman. This lack of understanding contributes to relatively high rates of missing data or misclassification (Ridolfo et al. 2012).

These types of problematic response patterns can be contrasted with those found among lesbian, gay, bisexual, and transgender (LGBT) respondents. While the problematic response patterns for non-minority respondents center on the lack of a salient

TABLE 7.1 Comparison of 2002 and 2006 NSFG Sexual Identity Measures

	Heterosexual	Homosexual	Bisexual	Something else?	Do not Know/Refused
	Heterosexual or straight	Homosexual, gay, (or lesbian)			
NSFG 2002–2003	89.6	1.7	2.4	4.2	2.0
NSFG 2006–2008	94.5	1.3	2.5	0.4	1.2

sexual identity, problematic response patterns for LGBT respondents are rooted within the complex process of negotiating and forming a sexual identity. The problematic response patterns found among LGBT respondents, then, relate to shifting sexual identities and use of non-traditional categories (e.g., queer, same-gender-loving), and for transgender respondents, the complex intersection between gender and sexuality (for more detailed discussion, see Ridolfo et al. 2012). Regarding the implication of question design, the contrast of problematic response patterns suggest that potential design solutions may be at odds for the two groups; while simplifying the question and providing concrete definitions related to sexual behavior and attraction may be the best for non-minority respondents, this solution would likely create more response problems for LGBT respondents. Previous cognitive interviewing work, however, has shown the importance of utilizing categories that respondents use in their everyday lives to describe themselves—a solution that is beneficial for both minority and non-minority respondents. As opposed to the more abstract, scientific labels (i.e., "homosexual" and "heterosexual") which respondents do not always understand and do not use to describe themselves, using the terms "straight," "lesbian," and "gay" does indeed improve question performance for many respondents of both populations.

It is impossible to know the extent to which misclassification occurs in survey data. In addition, it is impossible to determine the extent to which misclassification would be improved with the addition of the more meaningful categories. However, as previous work (Ridolfo et al. 2012) has shown, it is possible to glean insight by examining those cases that fall into the missing categories, specifically, the respondents who refused or answered "do not know" or "something else." Table 7.1, which compares the 2002 NSFG and 2006 NSFG[4] survey data, illustrates that survey data collected using the more abstract labels are associated with higher rates of "something else," "refused," and "do not know" responses. In the 2002 NSFG, in which respondents were only asked about being heterosexual, homosexual, bisexual,

[4]The sexual identity question used in the 2002 NSFG was: Do you think of yourself as heterosexual, homosexual, bisexual, or something else? The sexual identity question used in the 2006 NSFG was: Do you think of yourself as heterosexual or straight, homosexual or gay (or lesbian), bisexual, or something else?

or something else, 6.2% of the sample fell into the missing categories. With the simple addition of the terms "straight," "gay," and "lesbian," (categories that respondents are more likely to use in their everyday lives) missing rates fell to 1.6%.

Most noteworthy, those missing cases in 2002 did not occur randomly. As illustrated in Table 7.2, those respondents with lower education were more likely to provide "something else," "refused," and "do not know" responses. In 2002, 14.4% of women with less than a high school education, in comparison to 2.1% of those with more than a high school diploma, were coded as missing on this variable. The differential drops significantly in the 2006 data—to 3.8% with less than a high school diploma compared to 1.0% for those with more than a high school degree. While women have higher rates of missing data, it is important to note that the same pattern is found for men.

By and large, as illustrated in Tables 7.1 and 7.2, the 2006 NSFG questions for sexual identity represents a marked improvement from the 2002 questions. However, response problems remain in the 2006 questions; the Spanish version of the questionnaire provided no translation for the word "straight" because there is no comparable word in Spanish. Interestingly, as shown in Table 7.3, while the rates of missing data decreased in 2006 for English interviews, the rate of missing data for Spanish language respondents continues to be relatively high at 8.9% and 9.3% for Spanish-speaking men and women, respectively.

Previous cognitive interviewing findings also revealed that the word "straight" is not always understood as intended among English-speaking respondents, who interpret the word to mean "straight-laced." For those respondents who also believe the word "heterosexual" means being gay, simply inserting the word "straight" does not alleviate problems with misclassification. Similarly, addition of the word "straight" does not clarify the word "bisexual" for respondents who believe the term implies heterosexuality.

7.2.3 Development of an Improved Sexual Identity Question

In designing a new question, the 2006 NSFG version was used as a point of departure because it was regarded as the best performing question to date on a survey. In order to improve upon the 2006 NSFG version, it was determined that the goals for the new question would be to: (1) reduce misclassification of non-sexual minority respondents, (2) reduce rates of "do not know" and "something else," and (3) be able to sort non-minority from minority sexual identity cases for those respondents who respond "something else." Given these goals, revisions were based on findings from previous studies presented above. The revised question and the follow-up questions are presented below in Figure 7.1. Revisions were based on the following design principles: (1) use labels that respondents use to refer to themselves, (2) do not use labels that some respondents do not understand—particularly if those misunderstood terms are not required by any other group of respondents, and (3) use follow-up questions to meaningfully categorize those respondents answering "something else" and "do not know."

TABLE 7.2 Distribution of Missing Data by Education in the 2002 and 2006–2008 NSFG

	Men			Women		
Missing data	Less than high school	High school	More than high school	Less than high school	High school	More than high school
2002	11.4% ($n = 1361$)	8.0% ($n = 1505$)	2.1% ($n = 2055$)	14.4% ($n = 1702$)	7.9% ($n = 2167$)	2.1% ($n = 3767$)
2006	3.1% ($n = 1883$)	1.6% ($n = 1590$)	0.7% ($n = 2637$)	3.8% ($n = 1960$)	1.2% ($n = 1844$)	1.0% ($n = 3522$)
	2002: Rao–Scott Chi-square (2) = 63.47, $p < 0.05$ 2006: Rao–Scott Chi-square (2) = 17.14, $p < 0.05$			2002: Rao–Scott Chi-square (2) = 240.28, $p < 0.05$ 2006: Rao–Scott Chi-square (2) = 39.42, $p < 0.05$		

Note: Missing data = something else, refused and do not know responses

TABLE 7.3 Percentage of Missing Data by Language and Ethnicity in the 2002 and 2006–2008 NSFG

	Men			Women		
Missing data	Spanish Hispanic Interview	English Hispanic Interview	English Non-Hispanic Interview	Spanish Hispanic Interview	English Hispanic Interview	English Non-Hispanic Interview
2002	12.1% (n = 359)	10.6% (n = 763)	5.1% (n = 3793)	12.9% (n = 558)	9.5% (n = 1031)	5.6% (n = 6037)
2006	8.9% (n = 451)	1.3% (n = 947)	1.0% (n = 4708)	9.3% (n = 546)	1.2% (n = 1053)	1.1% (n = 5716)

Note: Missing data = something else, refused and do not know responses

Do you think of yourself as:

(For men:) Gay (For women:) Lesbian or gay

(For men:) Straight, that is, not gay (For women:) Straight, that is, not lesbian or gay

Bisexual

Something Else (*Go to A*)

Do not Know (*Go to B*)

A. (*If "something else" is selected*) By *something else*, do you mean that…

You are not straight, but identify with another label such as queer, trisexual, omnisexual, or pan-sexual

You are transgender, transsexual, or gender variant

You have not figured out your sexuality or are in the process of figuring it out

You do not think of yourself as having sexuality

You do not use labels to identify yourself

You made a mistake and did not mean to pick this answer

You mean something else (*Go to C*)

B. (*If "do not know" is selected*) You did not enter an answer for the question. That is because you:

You do not understand the words

You understand the words, but you have not figured out your sexuality or are in the process of figuring it out

You mean something else

C. (If "you mean something else" is selected)

What do you mean by something else? Please type in your answer

FIGURE 7.1 Revised sexual identity question and follow-up questions

The following discussion outlines the design principles and presents the rationale behind the new version.

1. *Use labels that respondents use to refer to themselves.*

   For minority group respondents, the word "gay" (and, for women, the phrase "lesbian and gay") was retained, while the word "homosexual" was dropped. As was discovered in the 386 interviews conducted in the QDRL since the year 2000, the most common labels used by sexual minorities to describe themselves are "gay," "lesbian," and "bisexual." In many cases, these respondents described the term "homosexual" as being overly scientific or connoting sexual deviance, not an affirmative label for gay and lesbian people. Besides "lesbian" or "gay," other less common words were used by sexual minority respondents to describe themselves, including "queer," "same-gender loving," "pansexual," and "trisexual." For these respondents who do not identify with the more commonly used categories, the category "something else" was provided and, from this question, they would be taken to a follow-up question that would allow them to clarify how they identify themselves. For the non-minority category label, the word "heterosexual" was replaced with the phrase, "not gay," because many non-gay respondents did not associate themselves with a sexual identity, but rather maintained a "not-me" identity.
2. *Do not use labels that some respondents do not understand—particularly if those misunderstood terms are not required by any other respondent.*

   The term "heterosexual" was dropped because many respondents, particularly those of lower socio-economic status, did not accurately understand the term. Significantly, in some cases, it was understood to mean "being gay." In no case did a respondent identify as heterosexual and require the heterosexual label; respondents either thought of themselves as not gay or as straight, or know that the word "straight" implied heterosexuality. The degree of confusion over the terminology along with the potential for misclassification suggested that the new question at least be tested without the word "heterosexual." The term "bisexual" was also often misunderstood by non-minority respondents with lower socioeconomic status. However, this term could not be replaced because it is a term that is commonly used by bisexual respondents. Furthermore, it was hoped that the preceding category, "straight, that is, not gay," would be clear enough that non-minority respondents would not erroneously choose "bisexual" thinking it implied heterosexuality.
3. *Use follow-up questions to meaningfully categorize those respondents answering "something else" and "do not know."*

   To further reduce the percentage missing, the revised question version would attempt to successfully classify these respondents into meaningful categories. While it is hoped that the use of "not gay" and the elimination of the terms "homosexual" and "heterosexual" would generate less cases of missing data, it is possible that some minority respondents would fall into "something else" or are still figuring out their sexual identity. A follow-up question was added, then, to sort these cases into the proper non-minority category.

One potentially controversial aspect of this revised question design is the response order. The general practice in question design is to place the most commonly chosen response option first. One reason for this is the primacy effect, that is, due to cognitive burden, respondents spend more time processing earlier options (Krosnick and Alwin, 1987). There are at least two reasons, however, why this would not apply to the question on sexual identity. The first, and most obvious, reason is that because the heterosexual option reads "straight, that is, not gay," and, therefore, requires the "gay" option to be listed previously. The second reason, and related to the first, is that respondents engage in satisficing, that is they look for the first option that is reasonable, even if not optimal (Tourangeau et al. 2000). By placing the non-minority response lower in the list it encourages respondents to more closely consider previous response options.

7.3 CASE STUDY: COGNITIVE INTERVIEWING EVALUATION OF THE NATIONAL HEALTH INTERVIEW SURVEY REVISED SEXUAL IDENTITY QUESTION

The methodological processes used to examine the performance of the newly revised sexual identity question were those outlined in this book. In this next section, study methods including recruitment and sampling, interviewing procedures, and data analysis will be described.

7.3.1 Recruitment and Respondent Demographics

To test the newly revised question, the QDRL conducted 139 cognitive interviews: 94 in English and 45 in Spanish. The interviews were conducted onsite at the QDRL interview lab in Hyattsville, Maryland as well as at several offsite locations in Washington, D.C. neighborhoods. English-speaking respondents were recruited through the QDRL database, newspaper advertising, flyers, and by word-of-mouth. Spanish-speaking respondents were recruited through flyers, by word-of-mouth, and with the assistance of several non-profit organizations catering to the Latino community.

Table 7.4 presents respondent demographics for the study. An attempt was made to capture a broad range of respondents but particular emphasis was placed on recruiting gay and lesbian respondents as well as a range of those reporting "something else," specifically, those who identify as transgender, queer, or who are still in the process of figuring out their sexuality.

7.3.2 Interviewing Procedures

Respondents were scheduled for specific interview times (with the exception of a few "drop-ins") and reported to a set location for their interview. Interviews lasted between 30 and 90 minutes with the typical interview lasting from 45 to 60 minutes. All interviews were audio recorded using both a cassette recorder as well as a sound-recording program on the computer. Respondents were asked to check an anonymous

TABLE 7.4 Respondent Demographics

Interviews completed:	139	
	Count	Percentage (%)
Gender		
Male	65	46.8
Female	66	47.5
More complicated	8	5.8
Sexual identity		
Straight	86	61.9
Gay or lesbian	24	17.3
Bisexual	9	6.5
Something else	19	13.7
Education		
Less than HS degree	23	16.5
High school degree/GED	38	27.3
Some college, no degree	22	15.8
Associates degree	17	12.2
Bachelors	21	15.1
Graduate school	17	12.2
Race		
White	32	23.0
Black	62	44.6
Indian American	7	5.0
Asian	4	2.9
Other	18	12.9
Latino	49	35.3
Language		
English	94	67.6
Spanish	45	32.4
Age		
Under 25	21	15.1
26–40	45	32.4
41–60	48	34.5
Over 60	16	11.5

consent form before the interview began and were also asked to give their oral consent once the taping began. At the conclusion of the interview all respondents were given $50 as remuneration.

Unlike other QDRL interviewing projects, the questionnaire for this project was administered using an audio-computer assisted self-interview (ACASI) system. (Although not relevant to the findings of this report, the ACASI system was also being tested as one piece of this overall project.) Respondents were asked to answer 8–10 demographic questions along with the sexual identity question using the ACASI system without any assistance from the interviewer. At the conclusion, respondents were asked each item and were then asked to explain their answer. Typical follow-up questions included, "How so?" and "Why do you say that?" If a respondent's answer seemed vague or unclear, the interviewer asked: "Can you give an example to describe what you are talking about?" Specifically for the sexual identity question, respondents were also asked how they typically referred to themselves and were also asked about other words (i.e., "heterosexual" and "homosexual") that were not appearing in the question. The culminating text from the interview related how respondents understood or interpreted each question and also outlined the types of experiences and behaviors respondents considered in providing an answer.

7.3.3 Data Analysis

Data from the interviews were analyzed using the five-tier analytic process described in Chapter 4. This analytic process simultaneously reduces data while moving from individual interview text to a conceptual understanding of question performance. After interviews were collected (the first step of analysis), and summaries were written to distill relevant information from the full interview text (Step 2). From those summaries, common themes were identified across all interviews (Step 3). Themes identified from interviews for this project included:

- The basis for respondents' answers, that is, whether they reported the social component of sexual identity, their sexual behavior, or attraction.
- The ways in which respondents understood the various sexual identity labels.
- The ways in which respondents conceptualized their own sexual identity.
- Response problems:
  - Whether or not respondents understood the various labels.
  - Whether respondents were able to place themselves within the provided response options.

Analysis was then conducted across groups (Step 4) to determine whether specific patterns were more likely to occur within a particular group, indicating whether a particular group was more likely to interpret or process the question differently. Differences across groups could indicate incomparability and a sign of bias in resulting survey data. For this study, the relevant groups for comparison were sexual minorities and non-minorities as well as Spanish speakers and English speakers.

Finally, analyses from the previous steps were brought together to provide a coherent understanding of question performance, including difficulties experienced by respondents when attempting to answer the question, the phenomena captured by the question and any differences across groups. As discussed in Chapter 6, individual respondent quotations or examples are solely presented to illustrate the larger conceptual themes that derive from the five-tier analysis. In addition, examples are intended to provide detail and realism that reveal respondents' perspectives within the context of their own lives. They are not anecdotal stories separate from or outside of the analytic process. From this type of analysis, one that portrays question performance, an evaluation of the question can be made. That is, whether or not the question operates as it is intended and, if not, whether that performance is acceptable. The final step of the analysis is presented in Section 7.4.

To facilitate the analytic process as well to generate an audit trail of data reduction, Q-Notes software, an application developed by the National Center for Health Statistics for the specific use of analyzing cognitive interviews, was utilized.

7.4 CASE STUDY FINDINGS

This section contains the findings of the five-tier analysis. First, an overall summary of the question performance will be described. The remainder is organized by the themes developed in Stage 3 of analysis, specifically: (1) basis of respondents' answers, (2) response problems, (3) interpretation of the terms "heterosexual," "gay," "lesbian," "homosexual," "bisexual," and "something else."

7.4.1 Summary of Question Performance

In comparison to previous versions of the sexual identity question (including the 2006 NSFG version), data from the cognitive interviews indicate that this newly developed version is an improvement. In all but 10 of the 139 interviews, respondents selected the response category that best reflected their described sexual identity. That is, respondents' answers were based on the ways in which they conceptualize their own sexuality. This was true for all age and socio-economic groups. Notably, almost all heterosexual respondents opted for the "straight, that is, not gay," response option with no difficulty. By and large the question performed well, that is, respondents' answers were consistent with their described identity, and only once was a minor change made to the question. The Spanish translation, however, received a relatively significant alteration.

The presence of the "something else" category along with the follow-up question also proved to be a successful revision. All respondents who opted for this category were able to effectively classify themselves within one of the provided options. Unlike previous versions of the question, none of these respondents were heterosexual; non-minority respondents answered by selecting the "straight, that is, not gay" category. Thus, it is believed that the revision of the heterosexual category addresses the missing data problem, including heterosexuals choosing the "something else" category. It should also be noted that only one Spanish-speaking respondent selected the "do not know" option because she was not familiar with the terminology. The "something else" option was most frequently chosen by transgender respondents, who then selected the transgender option in the follow-up question. Other respondents who selected the "something else" option included those respondents who identify as queer, do not use labels to identify themselves, have not figured out their sexuality, or do not consider themselves to have a sexuality.

7.4.2 Basis of Respondents' Answers

Essentially all respondents answered the question based on their conception of self. That is, their answer was based on their sexual identity and not behavior or attraction. For sexual minorities, however, answers were based on a variety of social dimensions. These dimensions included membership in a larger community, political activism, various personality characteristics, and relationship status. What is true of all of these factors, however, is that they are all various mechanisms through which respondents make sense of their sexual identity. In addition, consistent with previous findings, heterosexual respondents did not typically identify within a sexual identity category, instead thinking of themselves as "not gay." Figure 7.2 presents a conceptual map of the constructs captured.

One way in which respondents framed their sexual identity was through membership in a larger community. Many of the respondents saw themselves as members of a larger sociopolitical group, and they conceptualized their identity based on an affiliation with a larger LGBT community. One respondent, for example, said that they define gay simply as "the whole community." Another mechanism by which respondents informed their sense of self was through what they perceived to be political activism. There was a clear theme among many of the minority-identified respondents that their sexual identity was strongly tied to a sense of political activism.

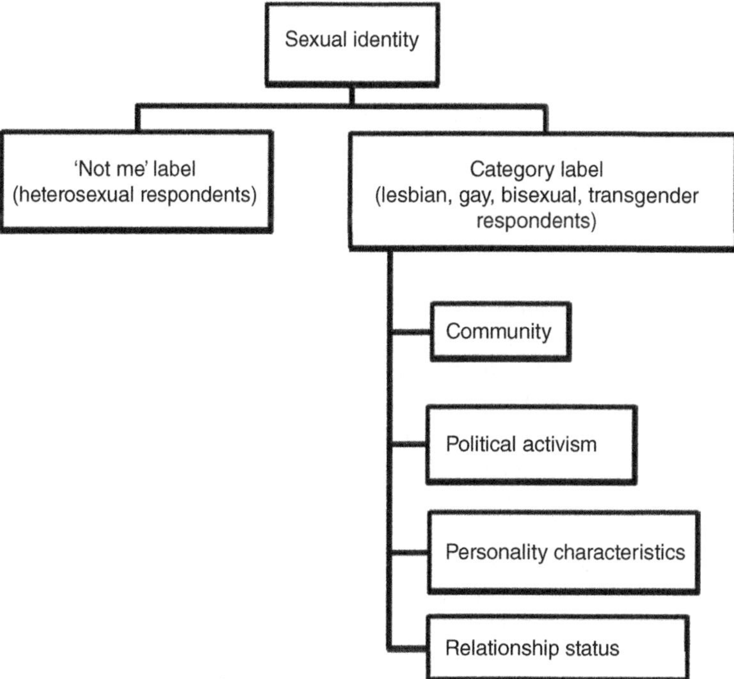

FIGURE 7.2 Conceptual map of sexual identity constructs

In a culture where homosexuality has been and continues to be heavily politicized, this sort of activist affiliation is seen to be a logical base for identity development. One respondent, for example, who identified as "something else" said that they do not really like to use labels but that they feel that they should do so in order to educate people.

Several of the respondents viewed certain personality traits as expressions of their (and others') identity as a sexual minority. One male respondent who identifies as gay, for example, made sense of his identity based on his perception of characteristics he finds to be inherent to gay people. He said that to be gay means to be happy, living a certain lifestyle that involves "being free, ecstatic, dramatic, full of zest and flavor." He went on to mention all of the artistic gifts that gay people have been given. He further noted that it had nothing to do with sex, as he has not had sexual relations in 5 years, yet he still identifies as gay. Another respondent said that to be gay means that he cannot think like a straight person. He said that straight people are more closed minded and focused on machismo while gay people are more open minded.

The sex of one's relationship partner was another mechanism by which some respondents made sense of their sexual identity. One female respondent, who identifies as "something else," for example, is currently in a relationship with a man but has been in relationships with women before. She said at the time of her relationship with a self-identified lesbian, she identified herself as "Maria-sexual," based on the name of her partner. She makes sense of her identity not based on behavior or attraction but rather based on the relationship that she is in at the time.

Transgender people often have a difficult time fitting into either the heterosexual or the lesbian/gay/bisexual community, although they feel a greater affinity for the latter. For this reason, many transgender respondents referred to the gay community in broader, more encompassing terms than lesbian/gay/bisexual or heterosexual respondents. Thus, a number of transgender respondents conceived of the term "gay" as both an individual identity as well as an umbrella term for a larger community of sexual minorities (the exact composition of that community varied among respondents). One transgender respondent said that although gay can specifically refer to a man who is masculine, it can also be used to refer to "the whole community." Another transgender respondent wanted to choose the term transgender but since it was not available chose gay because she felt that this was the closest option for her since it would include her in the LGBT community. Another transgender respondent said that she thinks of the term gay as being in the middle of a big circle of other terms like bisexual and transsexual and that "gay" is the word used to describe all of these things. She said that "gay" is the generic word used to describe all of these other terms, but that it is not specific enough and she would not identify this way. Instead, she identifies specifically as transsexual.

As was found in previous studies, many non-LGBT respondents did not possess salient sexual identities. In addition, as in previous studies, for these respondents it was not so much an association with a particular sexual identity that mattered as it was a disassociation from a gay identity. When asked about their sexual identity, many respondents simply said that they are "not gay." During probing a number of respondents indicated that they chose this option specifically because it said "not gay" and that this is what made the question easy for them to answer. One respondent, for example, felt that it was insulting to gay people to call oneself straight—"you are just not gay" she noted. Another respondent was asked if she would use the word "straight" to describe herself in her everyday interactions. After pausing for a moment the respondent answered, "I would just say that I am not involved in a gay relationship." She went on to say "I do not know why they use that word.... cuz really to me the word is "not gay." I do not know why people define it as straight and gay." Another respondent said that he was confused by the category "straight," but when he saw "that is, not gay," he knew immediately which category applied to him. Another respondent who identified as straight said that to her this meant that she "does not mess around or do things out of the ordinary." Another respondent said plainly that to be straight means she "do not act like they do."

7.4.3 Cases of Response Problems

Of the 10 respondents who did not answer according to their described sexual identity (and which could be considered error), three were sexual minority respondents and the other seven were Spanish-speaking respondents. Of the sexual minority respondents, two interpreted the question as a behavior question as opposed to an identity question and, consequently, answered bisexual. One woman, for example, who identifies as "queer," answered bisexual because she surmised that a health survey must be asking about her behavior, not her self-conceptualization. The other respondent not basing his answer on identity was transgender and answered according to the clinical records where his gender transitioning occurred, which was bisexual. While these cases do represent what would be considered error, it was deemed imprudent to make a revision to the question because any "fix" would likely generate other types of error.

The seven Spanish-speaking respondents who answered incorrectly were respondents who did not understand the word "gay," but were more familiar with the term "heterosexual." Since the word "gay" (along with the term "straight") is also an English-derived term, some of the Latino respondents were unable to make sense of the phrase "no es gay." For these respondents, absence of the term "heterosexual" generated more (as opposed to less in comparison to their English-speaking counterparts) response problems. For example, one Latino who answered "something else" later revealed that he is heterosexual but that he did not see that option listed for this question. Similarly, a Latina respondent answered "bisexual," but during probing revealed that, because she had to think very quickly and did not see the option for "heterosexual," chose "bisexual."

In addition, for Spanish-speaking respondents, because the word "heterosexual" was not listed, other terms, specifically "bisexual" and "lesbiana o gay," were misinterpreted. For example, one Latina who answered "bisexual" explained during probing that bisexuals are those who only sleep with men. Realizing her mistake she said "oh no! Bisexual means that they have sex with both men and women. I am heterosexual!" She went on to say that the response categories did not include the option she was looking for—heterosexual. Another Spanish-speaking respondent answered "lesbian or gay" because he was not sure what the word is for men who only like women. He could not remember if it was bisexual or heterosexual so he just chose the first response category listed. To resolve this response problem, the Spanish translation was modified shortly after these Spanish interviews, and it is believed that this modification will minimize, if not eliminate, these instances. It should be noted that none of the Spanish-speaking respondents had difficulty selecting the response category that best reflected their sexual identity after the word "heterosexual" was added.

7.4.4 Interpretation of Categories

"*Heterosexual.*" Perhaps most controversial about the revised question in comparison to the previous questions about sexual identity is the absence of the term "heterosexual" as a response option. For English interviews, no evidence was found to suggest the presence of response error or any response difficulty because the word "heterosexual" was not listed. This was true for all English-speaking demographic groups across heterosexuals. Even those respondents who said that they used the word "heterosexual" to self-identify were also familiar with the word "straight" or related to the concept "not gay." In no case did an English-speaking respondent indicate that they did not know how to answer because the word "heterosexual" was not there.

Consistent with previous studies, in the follow-up probing, it was found that many lower socio-economic non-minority respondents either did not know or misunderstood the term "heterosexual." For example, when asked what heterosexual meant, one English-speaking respondent said, "Who?," and then asked the interviewer what that word meant and how to pronounce it. Another female respondent noted that she was familiar with the term heterosexual but was not entirely sure what it meant. A number of respondents confused the term "heterosexual" with being homosexual and with being bisexual. For example, when asked what heterosexual meant one respondent answered that "it means men who like men." One female respondent explained that heterosexual means you can go with both men and women. Another respondent said that it is "somebody who goes both ways." Yet another respondent pointedly replied that "heterosexual means the same thing as bisexual."

Indeed, many of those who knew the definition of heterosexual remained unsure. When asked why he chose the answer he did, one respondent said it was because he identifies as "heterosexual or as someone who only likes women, unless I am wrong about the definition of heterosexual." Another respondent who was also unsure said that they were fairly confident it meant the same thing as straight but they were not totally sure about that. This last respondent emphasized the point that even for those respondents who might know the term "heterosexual," the use of more common language is a more guaranteed way to ensure respondent comprehension of response options.

Although sexual minorities tended to be more familiar with sexual-identity-related terms, there were instances, particularly related to the word "heterosexual," when they were not. One lesbian, for example, said that the word she would use for someone who likes the opposite sex is straight. When probed whether there was another word for this she said "I think the word is heterosexual, but maybe it is homosexual." She said that either way it did not matter to her because these words are basically for people who "deal with" the opposite sex.

For English-speakers, even among those who knew the term "heterosexual," there was still a clear preference for the word "straight." Several respondents noted that the term "heterosexual" (and on occasion, but not always, the term "homosexual") is a very scientific term and not what they use in everyday language. One respondent noted that he thought he had heard the term "heterosexual" in science class. An English-speaking male responded that he uses the word "straight" to describe himself normally and only uses the term heterosexual at school and when asked directly if he is a heterosexual or not. Most importantly, even among those who do use the word "heterosexual," straight was also understood.

In sum, the reasons for omitting the word "heterosexual" in a response option of the English version of the question are three-fold: (1) it is not the word that most people use in their everyday speech, (2) it is not required, as people understand the word "straight," and (3) many people are confused, do not understand, or misunderstand the word "heterosexual." The word "straight," although considered by some respondents to be slang, was understood by all English-speaking respondents and, equally as important, understood to mean what is implied by a heterosexual identity. The usage of the word "straight" and the removal of the word "heterosexual" in combination with the phrase "not gay," therefore, were found to greatly reduce conceptual confusion among respondents.

As previously noted, the above findings did not hold true for Spanish-speaking respondents. Because there is no word for "straight" in Spanish (although many Spanish speakers who had been living in the United States for a while were familiar with this term), the option simply read "no es gay." "No es gay" was not clear because the term "gay" is also an English term that is not always understood (as these few cases illustrate). For Spanish speakers, the term "heterosexual" was found to be much more commonly used and understood. That is, as far as usage and familiarity, the term "heterosexual" in Spanish is comparable to the term "straight" in English. For example, two respondents noted that they would have chosen "heterosexual" had they seen this option but since they did not see it instead chose "something else" and "bisexual."

Even among Spanish-speaking heterosexuals who did not have problems selecting the response option that best reflected their sexual identity, there was a strong sentiment that the presence of the word "heterosexual" would have made the question easier to answer. One respondent, for example, when asked how she understood "no lesbiana o gay" said "this is maybe where heterosexual goes?" Another respondent when asked what other words he would use to describe "no es gay" said that heterosexual was the most common word used for this. Yet another said that she found "no es gay" to be confusing and instead would have chosen the words "heterosexual" or "straight." Our data suggests that, although not always used, the term "heterosexual" is more commonly used among Spanish speakers. In response to this finding, the response option was changed from "no es gay" to "heterosexual, o sea, no es gay." This was then tested on 18 respondents, none of which had error or response difficulty.

"Gay," "Lesbian," and *"Homosexual."* As revealed in the comparison of the 2002 and 2006 NSFG data, the addition of the term "gay" appears to increase conceptual clarity among respondents. Because this is the word used most commonly by both sexual minorities and non-sexual minorities, it was scarcely unknown or misunderstood. The term "lesbian" was also commonly understood by respondents with no cases of conceptual confusion among either English- or Spanish-speaking respondents. The term was generally understood to mean the same thing as gay with the exception of one respondent who reported that she uses the word lesbian to refer to herself but does not use the word "gay." For example, when a respondent who reported that she uses both the word "gay" as well as "lesbian" to describe herself was asked which she preferred, she responded "I would choose lesbian, but it is still the same." Alternatively, another female respondent said that she uses both gay and lesbian to refer to herself but that she has a slight preference for gay. Another respondent said that she defines herself as a lesbian but that the term gay would also apply to her since it is a broader term encompassing "both men and women who like the same gender." Thus, although there was variation in preference for the term gay or the term lesbian, there was no conceptual confusion created by the term "lesbian."

Evidence was found to not use the term "homosexual" in the response options. Like the term "heterosexual," "homosexual" was often misunderstood or not known by respondents. One English-speaking respondent, for example, knew the term gay but not the terms heterosexual or homosexual. Another female respondent explained that, to her, being homosexual means being attracted to the opposite sex. In addition, and like the term "heterosexual," even when the term "homosexual" was understood it was often seen as an overly clinical term or, unlike the term "heterosexual," seen in a pejorative light. One Spanish-speaking respondent noted that to refer to gays she uses the term "chicos gays" because the word "homosexual" is "stronger" and has a negative connotation. An English-speaking respondent said that he only hears the word "homosexual" used when speaking disparagingly of people, for example, with reference to a "homosexual agenda." Another gay male acknowledged that homosexual does not have any inherently bad meaning but that people "do not use it properly .... and they say it with disdain."

Some respondents acknowledged that they might use the word "homosexual" but only in certain circumstances. For example, one gay male respondent said that he might identify as homosexual to a foreigner "who might not know what gay means." Another said he uses the word "homosexual" "only in the context of jokes." Another context for the usage of the word homosexual seems to be generational. For example, one 88-year-old respondent said that her grandchildren always correct her when she uses the word homosexual and tell her that the word is just gay. This latter point illustrates the larger point that even when homosexual is the preferred word choice, respondents are still familiar with the term gay.

There were a small number of Spanish-speaking respondents who noted that the term "gay" might not always be understood by other Spanish-speaking respondents because it is an English word. One respondent, for example, said that he uses the term "gay" with his friends in the United States but "homosexual" with his friends back in El Salvador. This would be consistent with our finding that those who had lived in the United States longer were also more likely to understand the term "straight," another English language slang (although this term was not used on the Spanish version, it was still mentioned by several respondents). One reason for this is that each country has its own specific slang for gay people, most of which are fairly insulting. One respondent, for example, said that the lower class in his country use the terms "maricones" or "culeros" and only the upper class really uses the term gay. Although this potential source of error should be noted, none of the 45 Spanish language respondents were unable to select the sexual identity that best represented them because of the presence of the term "gay" (or the absence of the term "homosexual").

While a few confused the term gay with being heterosexual or bisexual, a fair number of non-minority respondents believed that the term "gay" meant taking on some characteristic of being transgender, that is acting, dressing, or taking on the characteristics of the opposite gender. One respondent, for example, talked about gays as "men who wear ladies' clothes." This was echoed by another respondent who said that gay men dress like females and wear bras and skirts. Another respondent defined being a lesbian as someone with the body of a woman but the attitudes of a man. Yet another respondent answered that "a gay" is a man who dresses like women and likes men while another said a gay person is someone "trying" to be male or female, especially men who "try to play a female role." This misunderstanding was also found among Spanish-speaking respondents. One heterosexual Spanish-speaking respondent, for example, said that gay means when a man wants to be a woman or to act like a woman. Another Spanish-speaking respondent said that gay men are biologically men but want to be women and are not well defined in their identity. However, these misunderstandings did not appear to impact respondents' ability to properly select the response option that best reflected their sexual identity.

To a certain extent, some heterosexual respondents—particularly Spanish-speaking respondents—conflated being gay with a dimension of gender identity. For example, one respondent said "if you are male, you are straight. If you are female, you are straight." Another noted that "I am normal. I am a woman. I am feminine" which expresses the concept of gender identity but the reference to "normal" also implies an association with a "not-me" identity. A number of other respondents answered this question by simply saying "soy hombre" (I am a man) or "soy mujer" (I am a woman). The underlying theme of these respondents can be summed up by one who said that heterosexuals "do not feel like a man one day and a woman the next."

"*Bisexual*." Although there was some confusion over the meaning of the term "bisexual," it did not lead to response error problems because those people who did not know what the term meant did know the category with which they identified—that is, gay or not gay—and so knew for sure that bisexual was not the appropriate category. In the English-speaking cases where bisexual was chosen as response error, it was done not because of confusion over the term, but rather because these respondents thought the question was asking about behavior rather than identity (as discussed above).

That said, there were some respondents who did not know what the term "bisexual" means. One respondent, for example, said she had heard of the term but added "I do not quite understand what it means." Even many respondents who knew the meaning of the word "bisexual" still had definitions typically rooted in being gay or heterosexual. For example, one respondent said that being bisexual meant being "heterosexual and attracted to the same sex." This respondent started with an understanding of heterosexual and then built from it.

Other respondents confused the meaning of bisexual with either gay, heterosexual, or transgender. One respondent, for example, said that it was just another term for gay—"sounds like the same thing to me," he said. Another verified that it meant the same to her as heterosexual. One male respondent who identifies as gay but is married to a woman said that the word bisexual is just a "cover word" for people who think the word gay means something bad. Another respondent said that bisexual is either someone who watches a couple have sex or a woman who enjoys sex with men and women but they are not certain which one. Another respondent said that a bisexual person tries to be "a woman and a man at the same time."

Even among those who understood the general concept of bisexuality, there was still sometimes confusion over its precise meaning. One transgender respondent, for example, revealed that although he has sex with both men and women, he does not consider himself bisexual because he thinks bisexual means that half the time you are attracted to men and half the time you are attracted to women whereas he is attracted to women 80% of the time and men only 20% of the time.

Some respondents knew the concept of bisexuality, but not the word. One female heterosexual, for example, said that you can like a man and a woman at the same time but she was not sure what the word to describe this would be. An elderly female respondent seemed to understand the concept but not be familiar with the word. She said that someone is either gay or not gay (reinforcing the earlier point of the not-me identity) but that someone might be somewhere in the middle. She assumed, however, that this person would then select "do not know."

Confusion over the term bisexuality was also found among Spanish-speaking respondents. One such respondent said that bisexual meant someone who likes women but also "likes gays." Another Spanish respondent said that bisexuals have a personal conflict on how to define themselves. Another, unable to clearly articulate a definition, could only say that a bisexual is "someone who is a human being." This is further evidence that sexuality is conceived of differently among Spanish speakers.

Behavior seemed to be much more prevalent in respondents' conception of bisexuality. One heterosexual female, for example, said that bisexual means people who sleep with their own sex and the opposite sex. She said that unlike being gay, being bisexual necessarily involves sex. Another common response was that bisexual implies "going both ways," with follow-up references to sexual activity with both males and females. Along these lines one lesbian respondent said that bisexual means "when you do not know which sex you want to be with and you just take them both."

"*Something Else*." The response option for "something else" was well understood by those who identified as something else. Many transgender respondents, for example, selected something else on the basis of their transgender identity. Several of the trans respondents noted that the first thing they looked for was a "transgender" response option[5] but when they did not find this option, these respondents then chose "something else" assuming that that is what it meant. There were also respondents who identify as queer, do not use labels to identify themselves, and are asexual—all sub-options of the "something else" response category—who were also able to accurately select this category as the one that best reflected their sexual identity.

Even many of the non-transgender respondents felt that "something else" implied some variation of an understanding of transgender. One respondent, for example, said that something else is for those people who do not know what they want to be—male or female—and that they have not found their sexuality yet. Another respondent felt that maybe it was for people who did not want to openly identify as gay or who were transgender or "lost" and do not really know what they are. Others noted that it was a category for people who are not a lesbian or a homosexual. A gay male respondent said that "there are so many letters now" and so it gives people a chance to pick something different.

Some respondents, especially those who did not identify as "something else," had varying initial conceptions of what the "something else" category could possibly mean or simply had no idea what it might imply. A heterosexual female, for example, said that something else made no sense to her because either you are straight or you are not. Another female respondent said that something else could be a hermaphrodite. She said that she knew a couple of hermaphrodites and that these are people born "with both sexes, both organs," and then their parents decide if they want to raise them as a boy or a girl. Another respondent said it was for someone who does not know if they like men or women and is the same as the "do not know" option.

7.4.5 Study Conclusions

Overall, our analysis of the 139 cognitive interviews leads to at least four main conclusions:

First, the absence of the word "heterosexual" on the English language question is helpful to reduce response difficulty. It is important to use common vernacular in order to reduce conceptual confusion. Thus, while the absence of the term "heterosexual" did not lead to any confusion among respondents in any demographic, its presence did.

Second, the presence of the word "heterosexual" on the Spanish language question helps respondents make sense of other response categories. Since there is no conceptual translation for the word "straight" in Spanish the presence of "heterosexual," a word more commonly used by Spanish speakers than English ones, is useful to provide context not only for this option but for the others as well.

[5]This was certainly not the case for all trans respondents as some chose "gay or lesbian" or "straight, that is not gay" without debate. Therefore, it is critical to understand that this question is a sexual identity question only; it should not be used as a gender identity question.

Third, for many heterosexuals the concept of sexual identity is not salient. They do not so much identify with being heterosexual as they disidentify with being gay. To this end, the addition of "that is, not gay" was useful in helping these respondents select the optimal response category.

Finally, due to the presence of the "not gay" wording, it is necessary to put this response category lower than the "gay" category. This is not only logically more correct, it also encourages respondents to more carefully consider previous response options.

7.5 CONCLUSION

The case study presented in this chapter provided an example of the methodological approach outlined in this book. In doing so, the chapter illustrated how a cognitive interviewing report should be written in its entirety, including a methods section that details the methodological steps implemented in the study as well as findings that are meaningful and tied to the original data source. Documenting studies in this way demonstrates the credibility of study findings. Importantly, this study utilized findings from existing cognitive interviewing studies. By drawing on previous studies, cognitive interviewing studies can be confirmed and, significantly, can improve upon existing question design.

8 Analysis Software for Cognitive Interviewing Studies: Q-Notes and Q-Bank

JUSTIN MEZETIN and MEREDITH MASSEY
National Center for Health Statistics

8.1 INTRODUCTION

It can be difficult to manage the large quantities of data that are collected during a cognitive interviewing project. The complex task of keeping data organized and accessible can be daunting for researchers. The process of managing data and publicizing results can be facilitated by the use of technology. Specifically, qualitative analysis software can assist throughout the analytic steps of a cognitive interviewing project, and there are many qualitative software programs available to researchers. This chapter discusses the Q-Suite resources for cognitive interviewing research. Q-Suite contains the applications of Q-Notes and Q-Bank. Q-Notes is an online qualitative software program developed specifically for cognitive interviewing analysis, and Q-Bank is an online resource for dissemination of cognitive interviewing findings. Q-Notes and Q-Bank are products of the Questionnaire Design Research Laboratory (QDRL) at the National Center for Health Statistics and were developed in order to assist with the collection, organization, analysis, and dissemination of cognitive interviewing data.

More than simply a database management system, Q-Notes has the functionality to allow researchers to conduct the various tiers of analysis in a systematic, transparent fashion. Q-Notes allows for easy and accessible data entry by centralizing the data entry process. All interview data are located in a single convenient project location rather than separate files and folders. This centralization allows for consistency across interviews because all data are recorded in a uniform format. Q-Notes helps researchers to conduct analysis according to the five-step pyramid approach outlined in Chapter 4. Specifically, Q-Notes is uniquely designed to help researchers analyze

Cognitive Interviewing Methodology, First Edition.
Edited by Kristen Miller, Stephanie Willson, Valerie Chepp, and José-Luis Padilla.

data when conducting interviews, summarizing interview notes, developing thematic schema, and drawing conclusions about question performance.

Q-Notes enhances transparency in the research process, providing users with an audit trail of interview notes, coded themes, respondent demographics, and frequency counts of the sub-group and cross-group comparisons that inform the thematic schema. This transparency enhances the reliability of the cognitive interviewing findings, as researchers coming in at any phase of analysis can use this audit trail to reconstruct the analysis in order to draw similar conclusions or, alternatively, build upon existing ones. Taking a step-by-step approach, this chapter illustrates how Q-Notes facilitates each analytic stage of a cognitive interviewing study, thereby allowing for a more efficient, systematic, and transparent research process. The chapter concludes by demonstrating how Q-Bank is used to disseminate findings of cognitive interviewing projects.

8.2 Q-NOTES ANALYSIS FEATURES

Q-Notes offers many features that can facilitate analysis in cognitive interviewing studies throughout the analysis process. The first level of analysis, conducting the interview, is done outside of Q-Notes using a question-by-question interview guide generated through Q-Notes. The second level of analysis, synthesizing interview text into detailed summaries, is done directly in Q-Notes. After all summaries are entered, an analyst can then use Q-Notes to compare narratives across respondents to uncover thematic schema. Q-Notes also has tools that can assist in the comparison of respondents across sub-groups in order to develop advanced schema. After completing these levels of analysis, a researcher can draw conclusions about question performance. Q-Notes provides an audit trail detailing exactly how those conclusions were reached.

Drawing upon a sample project for demonstration purposes, this section provides a step-by-step explanation of how the features of Q-Notes can help facilitate each stage of analysis. Although it factors more prominently in some analytic steps than in others, Q-Notes can be useful to researchers during each step in the analytic process. The sample project is based on an actual project that was conducted to test child disability questions for use in international surveys. Twenty-five respondents answered survey questions about their children's functional difficulties in several domains.[1]

8.2.1 Level 1: Conducting Interviews

The first level of analysis occurs when the cognitive interview is conducted. Q-Notes can be useful both before and after the interview takes place. Prior to the interview, Q-Notes provides the capability to create and print out an interview guide (also referred to as an interview protocol) to use during the interview. The interview

[1]This demonstration project is adapted from a real project; however, interview notes, respondent demographics, and other potentially identifying information have been changed to protect the privacy of study participants.

guide displays all the questions that will be tested in the cognitive interview, along with the response options. The guide also provides space for the researcher to write interview notes.

This feature of Q-Notes facilitates analysis at this "pre-interview" stage of the research process by ensuring that all interviewers ask the question exactly as written and that all answers can be properly recorded. If needed, instructions can also be added to indicate whether respondents should be skipped out of specific questions. For example, in Example 8.1, if the respondent answers "no difficulty," the interviewer knows to skip to Question 6. In particular, when multiple researchers are conducting interviews for the same project, the interview guide provides an open ended framework for all of the interviewers to follow. The interview guide includes a list of preselected themes for each question. Preselected themes included in the interview guide are based on researchers' pre-existing knowledge or hypotheses that should be of focus in the interviews.

Example 8.1 presents one question from the interview guide intended to perform as a children's disability measure: "Compared with children of the same age, does (he/she) have difficulty with self-care such as feeding or dressing him/herself?" Here, the preselected themes focus on response processes that all interviewers should address in each interview.

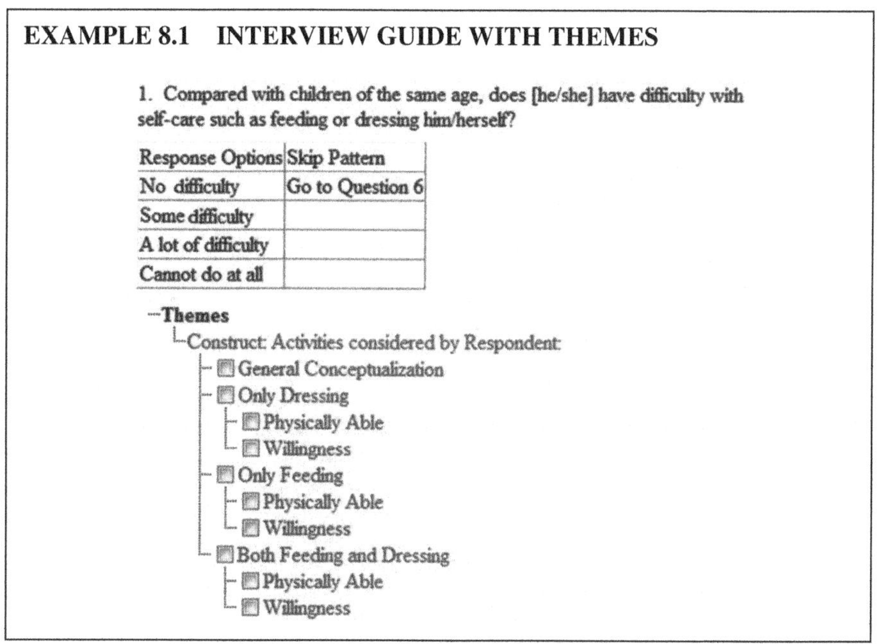

EXAMPLE 8.1 INTERVIEW GUIDE WITH THEMES

1. Compared with children of the same age, does [he/she] have difficulty with self-care such as feeding or dressing him/herself?

Response Options	Skip Pattern
No difficulty	Go to Question 6
Some difficulty	
A lot of difficulty	
Cannot do at all	

–Themes
- Construct: Activities considered by Respondent:
  - General Conceptualization
  - Only Dressing
    - Physically Able
    - Willingness
  - Only Feeding
    - Physically Able
    - Willingness
  - Both Feeding and Dressing
    - Physically Able
    - Willingness

The theme entitled "construct" pertains to the phenomena that respondents considered when formulating their answer. Previous studies revealed that respondents considered either their child's ability to dress, their ability to eat, or both dressing and eating. Previous studies also revealed that while some parents considered their child's physical ability to perform these activities (as was intended), other parents considered their child's ability to choose appropriate attire and whether they are "fussy eaters" (willingness). Including these predetermined concepts in the guide ensures that all interviewers collect this type of data when conducting interviews and that there will be little missing data. From this information, a schema can be more easily developed.

There are additional benefits to using the interview guide. This feature can be especially helpful when doing analysis for cognitive testing projects conducted in multiple languages. If multiple translations are being tested, an interviewer can print out an interview guide in the specific language used to conduct the interview. Moreover, use of the interview guide supports iterative testing approaches. That is, if researchers decide to change the question wording or any other information on the interview guide based on the ongoing analysis, this can be easily changed in Q-Notes. As multiple interviewers are testing on the same project, the interview guide serves as a central location for the most up-to-date changes. This helps to further ensure consistency across interviews.

At this stage of analysis, the researcher conducts the cognitive interview and can write their notes on the printed interview guide. It is a common practice for interviewers to have a transcript of the interview. This transcript can be in a written verbatim text, audio, and/or video format. The use of a transcript is not necessary, but it is a good supplement to the interviewer's notes and memory. The recording or the transcript, as the raw data, are retained for an audit trail and used in the next step of analysis.

8.2.2 Level 2: Summarizing Interview Notes

The second level of analysis occurs when the researcher develops a summary by synthesizing the raw data of an interview. Several features of Q-Notes help to facilitate this analytic stage. First, an individual respondent's answer to a survey question is recorded using a dropdown menu. Q-Notes also provides a simple interface for the interviewer to enter their interview summary for that question. If interviews have been video recorded, Q-Notes has the capability to organize interview data into tagged video clips that can be uploaded alongside a corresponding question. Thus, users can simultaneously see both the raw data and the detailed summary for a particular respondent. This allows analysts to check the accuracy and completeness of their summaries. Also, in this way, direct quotations can easily be pulled from the raw data for later use in analysis and reports.

In Example 8.2 from the sample project, the respondent's answer to the question Self-care_1 has been recorded as "some difficulty." The application assigns each interview with a respondent identification (in this example CD 14) as well as the interviewer name (MMassey). The interviewer, MMassey, has summarized the interview in the box provided. The summary provides relevant details from the interview that explain why the respondent answered "some difficulty." The interviewer has also included a side note about the respondent's interpretation of the question. This note can be used later by this particular researcher or other researchers who are working on the project. Notes may lead to the development of themes or may highlight potential response error. Next to the summary box is the tagged video clip of the segment of the interview that pertains to this question.

EXAMPLE 8.2 QUESTION SUMMARY WITH VIDEO

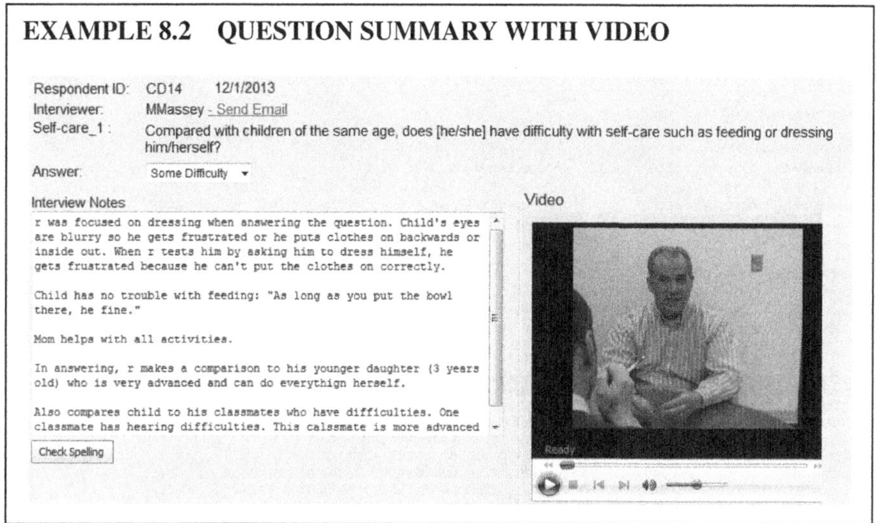

The Q-Notes interface facilitates collaboration between multiple users. While users may be in different locations, have differing levels of expertise, and may even be interviewing in different languages, the common interface ensures consistency in data format. The Q-Notes email function facilitates communication among users. For example, if an analyst has a question about a particular interview summary, he or she can send an email to that interviewer directly from the Q-Notes screen.

Once summaries for each question have been entered for a particular respondent, Q-Notes allows the researcher to view all of the data for that respondent in one screen. Details from other sections of the interview may be relevant to a particular question. From this screen, summary notes can be edited to reflect these details. Example 8.3 presents the characteristics and background information of respondent CD 14 as well as his answers and the interview summaries for the first three disability questions. There is a typo in the interview summary for the first question. This can easily be corrected by clicking on "Edit Notes." Reading through the interview details and the question summaries for this respondent, the researcher can see that this respondent's answers are heavily influenced by his child's difficulties with vision. For example, in the first question dealing with self-care, the parent discusses the child putting on clothes inside or backward, due to the child's blurred vision. In the third question, "Compared with children of the same age, how much does (he/she) worry or feel sad," the parent discusses how wearing an eye patch contributes to the child's frequent feelings of sadness. Seeing all of the responses for this respondent on one screen helps the researcher identify these patterns, which can guide later analysis.

EXAMPLE 8.3 WITHIN INTERVIEW ANALYSIS FEATURE

Within Interview Analysis

Select a respondent to view a respondent's answers and interviewer's notes for all questions.

Select Respondent: CD14 ▾ Print

Respondent Information

Respondent ID:	CD14	Interview Date:	12/1/2013
Interviewer:	MMassey	Place of Interview:	Laboratory
Country:		Language of Interview:	English
Gender:	Male	Age:	5
Marital Status:	Never Married		
Interview Summary Notes:	r was a young mother with 2 children. Focus of interview is older son, 5 years old. Child has a blurry eye and is supposed to wera a patch. Surgery was done, but did not correct the problem completely. r states that she does everything for the child and he is never out of her company. Attends a school for children with functional difficulties two days a week. r also has a 3 year old daughter who does not have any difficulties.		

Question	Notes	Answer		
Compared with children of the same age, does [he/she] have difficulty with self-care such as feeding or dressing him/herself?	r was focused on dressing when answering the question. Child's eyes are blurry so he gets frustrated or he puts clothes on backwards or inside out. When r tests him by asking him to dress himself, he gets frustrated because he can't put the clothes on correctly. Child has no trouble with feeding: "As long as you put the bowl there, he fine." Mom helps with all activities. In answering, r makes a comparison to her younger daughter (3 years old) who is very advanced and can do everythign herself. Also compares child to his classmates who have difficulties. One classmate has hearing difficulties. This calssmate is more advanced and does not get as frustrated or give up as often as her son.	Some Difficulty	Edit Notes	Themes
[If child uses a hearing aid] Does [he/she] have difficulty hearing when using his/her hearing aid(s)?	r says, "When he wants to play he will not listen to you. When he is watching his cartoons, he ignores." r says she has to scream at him to get him to respond. "It's not just his hearing it's his stubbornness too." Asked if r had taken child to a doctor to have hearing tested, says no. [Note: r seems to be addressing a listening issue rather than a hearing issue.]	Some Difficulty	Edit Notes	Themes
Compared with children of the same age, how much does (he /she) worry or feel sad?	The patch causes him a lot of trouble. He compares himself to other kids who don't wear a patch. His disability diminishes his feelings of confidence. He doesn't want to wear a patch in public. In general, he has a "regular level of sadness" but just lots of worry about his situation. r says this is the same level of worry as other kids at his school [for children with functional difficulties]. For example, some kids at school don't want to use their hearing aids. "He gets it from me. If I had to be in a big group and put a patch on my eye. I'd say lets just go home." r doesn't make him wear the patch much.	More	Edit Notes	Themes

8.2.3 Level 3: Comparing Across Respondents

During the third level of analysis, researchers compare across respondents in order to develop thematic schema. Example 8.4 shows the Across Interviews analysis function for the children's self-care question. This function displays the notes for all respondents for the particular question. Viewing all of the notes for a single question allows the researcher to begin to identify themes and patterns that appear across respondents. Themes may include items related to question interpretation, how respondents formulated their answers or response problems. For example, if multiple respondents asked for clarification of the question, that would be apparent in this screen.

EXAMPLE 8.4 ACROSS INTERVIEW ANALYSIS FEATURE

Select a Question: Self-care_1 ▾
Compared with children of the same age, does [he/she] have difficulty with self-care such as feeding or dressing him/herself?

Answers: ▾

☐ Project Themes
☐ Question Themes
☐ Demographics

☑ Show Question Answers
☑ Show Interview Notes
☐ Show Frequency Table

Output Results

RespondentID	Answer	Narrative
CD2	A lot of difficulty	Child is in a wheelchair. Cannot feed herself- food has to be served to her. On probing, r says child can feed self from a bowl with a spoon, but can't prepare or retrieve food because she's in a wheelchair. Child needs help in dressing. "If she dresses herself, the front portion may appear on the backside."
CD8	A lot of difficulty	Child has a lot of difficulty when it comes to eating as well as dressing herself. She is too slow when she's eating because she has weak coordination of herlimbs. She usually receives help from her family in eating. This is also why she has trouble dressing. Sometimes she dresses herself and it takes up to an hour to fully dress. Usually her family provides assistance.
CD6	A lot of difficulty	Food has to be served to the child. All food has to be of the proper consistency so the child does not choke. She only swallows and does not chew so the food can get stuck in her throat. r did not address child's difficulties dressing. r was only thinking of difficulties feeding when answering the question.
CD11	Cannot do at all	Child is dependent for all his daily living activities and physically unable to attend to these things. No comparison made to other children. r: "There's no comparison. He is completely dependent. I don't need to make a comparsion to know this."
CD13	No Difficulty	She is able to feed and dress herself with no difficulty. r says her daughter helps prepare food for family meals. She enjoys fashion and picks out her own clothes and has never had a problem attending to her hygeine.
CD7	No Difficulty	She can take a shower on her own. She can dress herself and has no trouble eating. She has a hard time tieing her shoes or using a knife (despite these difficulties r still answered no difficulty). When probed as to why she did not answer "some difficulty," r says that the child can take care of her daily needs so there is no difficulty.
CD10	No Difficulty	Child can prepare his own breakfast and eat it in the correct way. He keeps his clothes clean when eating. He can dress up well and has no trouble buttoning and unbuttoning his shirt without assistance.
CD3	No Difficulty	r said child doesn't have any difficulty. When he comes from school, he serves food by himself and eats it. He dresses and bathes and takes care of his hair by himself.
CD5	Some Difficulty	r explains that he has difficulty in dressing himself. He can't button his shirt or tie his tie or tie the laces of his shoes. Child is occupational therapy to help with these issues. OT has improved his abilites in this area. Child has trouble tucking his shirt into his pants and sometimes wears underpants front side back. Feeding: child has no trouble eating. He does not spill and keeps his hands clean.
CD14	Some Difficulty	r was focused on dressing when answering the question. Child's eyes are blurry so he gets frustrated or he puts clothes on backwards or inside out. When r tests him by asking him to dress himself, he gets frustrated because he can't put the clothes on correctly. Child has no trouble with feeding: "As long as you put the bowl there, he fine." Mom helps with all activities. In answering, r makes a comparison to her younger daughter (3 years old) who is very advanced and can do everythign herself. Also compares child to his classmates who have difficulties. One classmate has hearing difficulties. This calssmate is more

In Example 8.4, the researcher can quickly determine that, in formulating their answers, respondents considered one of four different activities: (1) only feeding, (2) only dressing, (3) both feeding and dressing, and (4) a general conceptualization of self-care. Because qualitative methods require flexibility, Q-Notes allows new themes to be added as they emerge. Further refinement of themes is possible through the creation of sub-themes and/or categories. For example, with further analysis, it was discovered that those respondents considering "only feeding" either thought about their child's physical ability to feed themselves, or their child's willingness to eat what they were given. Similarly, those respondents considering "only dressing" either thought about their child's physical ability to dress themselves, or their child's willingness to wear appropriate clothing. Figure 8.1 illustrates the various activities considered by parents as they formulated their answer.

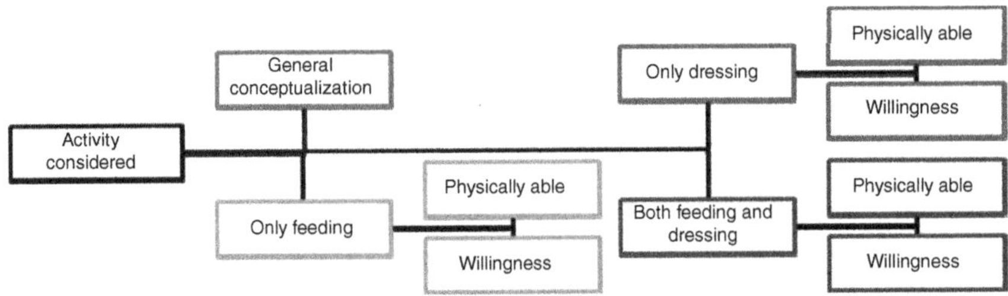

FIGURE 8.1 Thematic schema for self-care question

Once the thematic schema is defined, Q-Notes allows the researcher to code all respondent narratives accordingly. The common set of themes maintains consistency even when narratives are coded by different researchers. Further, when themes are selected in Q-Notes, a text box appears, allowing the analyst to enter direct quotes or an explanation for the coding selection. Sometimes arriving at a code is not a straightforward task and requires analysts' judgment. The text box allows other analysts to understand the rationale behind the selection, increasing communication and collaboration within the project. Example 8.5 shows how direct text from the summary is used to support the researcher's selection. This level of analysis is completed when all narratives have been coded, meaning the correct categories have been selected for all themes, for all questions.

EXAMPLE 8.5 DEVELOPING THEMATIC SCHEMA FROM THE DATA

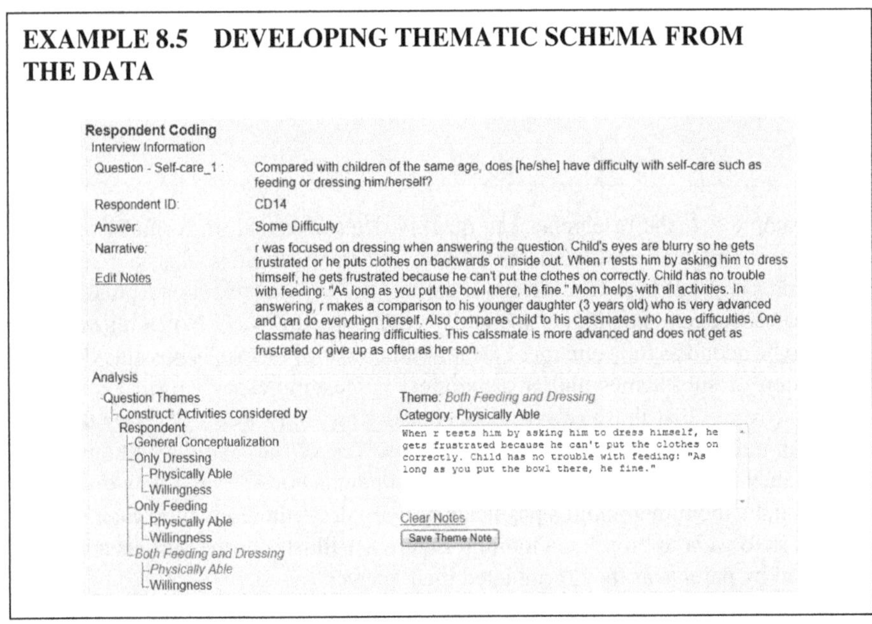

8.2.4 Level 4: Comparing Across Groups

The fourth level of analysis entails the development of advanced schema. In this stage of analysis, Q-Notes makes it easy to uncover intergroup relationships. Through systematic comparison across groups of respondents, it can be determined whether themes are more prevalent among specific groups of respondents. Groups can be defined in multiple ways, including demographic groups, common social experiences, or according to similar responses to survey questions. In all these scenarios, Q-Notes allows users to view, sort, filter, and group the project data in order to easily and efficiently identify relationships.

Two analytic functions in Q-Notes are especially helpful while comparing across groups, the Sub-Group Analysis and Comparative Analysis functions. The Sub-Group Analysis function is similar to the Question Level Analysis function and allows users to filter outputs according to specific response options or other criteria. This Sub-Group Analysis function displays the accompanying notes or frequency tables for a particular question.

Example 8.6 shows the search screen for the Sub-Group Analysis function. A researcher has selected the Self-care_1 question and wants to also see the age of the child as well as the language of the interview. Example 8.7 shows the results table that would display from the search conducted in Example 8.6. Using this function, the researcher can explore whether respondents answering about older children process questions differently than those with younger children. Similarly, the researcher can examine whether Spanish speakers interpreted questions differently than English speakers, which could indicate a possible translation problem.

EXAMPLE 8.6 SUB-GROUP ANALYSIS

Sub-Group Analysis

Select a Question: Self-care_1
Compared with children of the same age, does [he/she] have difficulty with self-care such as feeding or dressing him/herself?

Answers:

Project Themes
Question Themes
Demographics

Show Question Answers
Show Interview Notes
Show Frequency Table

Project Level Themes
View Main Themes View All Themes
Theme: disability status Category
Output: Notes All Categories

Demographics

Field	Output	Sort Order
Age	☑	
Country	☐	
Language	☑	
Gender	☐	
Interview Location	☐	
Marital Status	☐	

Output Results

EXAMPLE 8.7 SUB-GROUP ANALYSIS RESULT

RespondentID	Answer	Narrative	Age	Language
CD6	A lot of difficulty	Food has to be served to the child. All food has to be of the proper consistency so the child does not choke. She only swallows and does not chew so the food can get stuck in her throat. r did not address child's difficulties dressing. r was only thinking of difficulties feeding when answering the question.	6	English
CD8	A lot of difficulty	Child has a lot of difficulty when it comes to eating as well as dressing herself. She is too slow when she's eating because she has weak coordination of herlimbs. She usually receives help from her family in eating. This is also why she has trouble dressing. Sometimes she dresses herself and it takes up to an hour to fully dress. Usually her family provides assistance.	16	English
CD11	Cannot do at all	Child is dependent for all his daily living activities and physically unable to attend to these things. No comparison made to other children. r: "There's no comparison. He is completely dependent. I don't need to make a comparsion to know this."	11	English
CD7	No Difficulty	She can take a shower on her own. She can dress herself and has no trouble eating. She has a hard time tieing her shoes or using a knife (despite these difficulties r still answered no difficulty). When probed as to why she did not answer "some difficulty," r says that the child can take care of her daily needs so there is no difficulty.	17	English
CD10	No Difficulty	Child can prepare his own breakfast and eat it in the correct way. He keeps his clothes clean when eating. He can dress up well and has no trouble buttoning and unbuttoning his shirt without assistance.	12	English
CD3	No Difficulty	r said child doesn't have any difficulty. When he comes from school, he serves food by himself and eats it. He dresses and bathes and takes care of his hair by himself.	11	English
CD14	Some Difficulty	r was focused on dressing when answering the question. Child's eyes are blurry so he gets frustrated or he puts clothes on backwards or inside out. When r tests him by asking him to dress himself, he gets frustrated because he can't put the clothes on correctly. Child has no trouble with feeding: "As long as you put the bowl there, he fine." Mom helps with all activities. In answering, r makes a comparison to her younger daughter (3 years old) who is very advanced and can do everythign herself. Also compares child to his classmates who have difficulties. One classmate has hearing difficulties. This calssmate is more advanced and does not get as frustrated or give up as often as her son.	5	English
CD1	Some Difficulty	r's daughter has problems dressing herself. When r asked her to get ready this morning for school, the buttons on her shirt were mismatched and her daughter didn't even realize it. The other day daughter went to school with shoes on the wrong feet. No problems feeding herlself.	6	English
CD9	Some Difficulty	r says child is able to stay home alone. He can eat by himself, get ready and go out alone. When probed about why r answered "some difficulty" r says child can not tie shoe laces and does not remember to tuck his shirt into his trousers.	10	English
CD5	Some Difficulty	r explains that he has difficulty in dressing himself. He can't button his shirt or tie his tie or tie the laces of his shoes. Child is occupational therapy to help with these issues. OT has improved his abilites in this area. Child has trouble tucking his shirt into his pants and sometimes wears underpants front side back. Feeding: child has no trouble eating. He does not spill and keeps his hands clean	7	English

The second tool commonly used to compare across group is the Comparative Analysis function. This function provides a quick way of viewing and identifying relationships from a slightly broader perspective than the Sub-Group Analysis function. The Comparative Analysis tools allow users to create crosstabs of two different types of data. Q-Notes has four different data types:

- Questions: Which response options were chosen for the question.
- Question Themes: Which categories were chosen for the question theme.
- Demographics: Age, gender, country, marital status, or language.
- Project Themes/Interview Descriptors: Additional descriptors created for the project.

With this tool, subgroups can be defined based on those data types revealing relationships and patterns between the groups. In Example 8.8, the researcher has chosen to look at the relationship between the answer to the Self-Care_1 question and the "disability status" project theme. The "disability status" project theme identifies respondents who reported their child as disabled. By comparing these subgroups, the researcher can examine whether parents who identified their children as disabled differ from the responses of those who did not. The results of this search, presented in table format in the middle of the screen, allows the analyst to see numerical data patterns which can help guide the focus of the analysis.

The results presented in Example 8.8 reveal that those parents who identified their child as being disabled (disability = yes) most often chose "a lot of difficulty" which might be expected. However, those who did not identify their child as being disabled (disability status = no), most often chose "some difficulty" which may not be expected. Further investigation is necessary to understand why these parents chose "some difficulty." Patterns such as these are easily displayed, allowing the researcher to quickly see relationships in the data and to draw conclusions about these relationships.

EXAMPLE 8.8 COMPARATIVE ANALYSIS

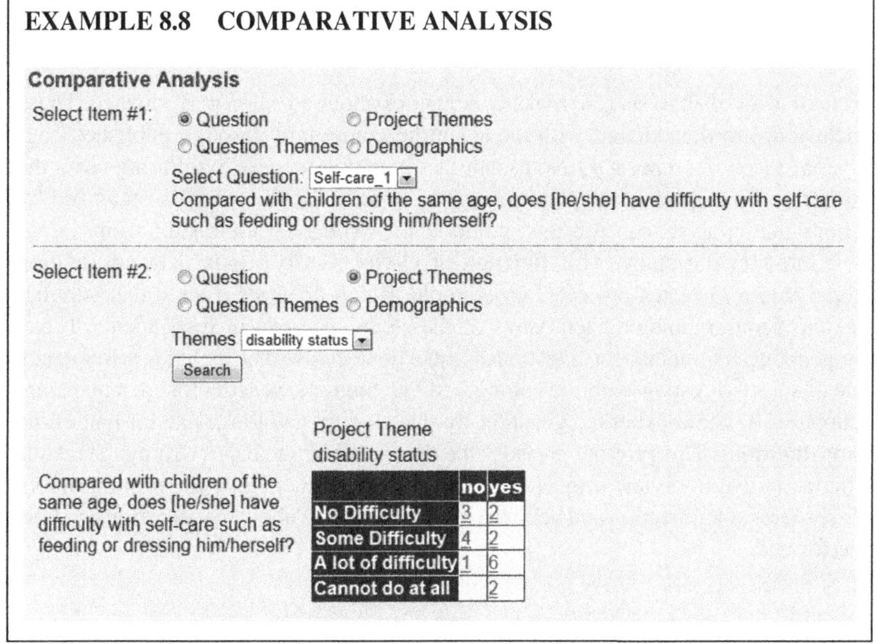

Comparative Analysis

Select Item #1: ◉ Question ○ Project Themes ○ Question Themes ○ Demographics

Select Question: Self-care_1

Compared with children of the same age, does [he/she] have difficulty with self-care such as feeding or dressing him/herself?

Select Item #2: ○ Question ◉ Project Themes ○ Question Themes ○ Demographics

Themes disability status

Search

Project Theme
disability status

Compared with children of the same age, does [he/she] have difficulty with self-care such as feeding or dressing him/herself?

	no	yes
No Difficulty	3	2
Some Difficulty	4	2
A lot of difficulty	1	6
Cannot do at all		2

Curiously, one respondent with a non-disabled child reported "a lot of difficulty." Clicking the number in any one of the cells displays the respondents that fall into that category as well as the summaries for those respondents. In this case, for example, the search for the respondent with a non-disabled child who reported "a lot of difficulty" was respondent CD17. Reading the interview notes for this particular case explains the false-positive response: the respondent based her answer by considering her child's willingness to eat the food that she is given (Example 8.9).

EXAMPLE 8.9 COMPARATIVE ANALYSIS RESULTS

Comparative Analysis

RespondentID	Narrative
CD17	When asked the question, the respondent sigh and shook her head. She said that she typically fights with her child about eating the food that she is given. She is very frustrated by this problem. Her child refuses to eat any type of vegetable and only certain types of fruit. She said that the only time that her child will eat what she is given is when they have pizza, macaroni and cheese, and hot dogs. She answered 'a lot of difficulty' as opposed to 'cannot do,' because on these occasion (when they have pizza, etc) the child will eat.

8.2.5 Level 5: Drawing Conclusions about Question Performance

The final level of analysis is to make conclusions about question performance. These conclusions are then shared with the scientific community through published cognitive interviewing reports. Q-Notes can facilitate this process by offering users the option to print a generic template for a final report. This template is organized by sections that analysts can use as a guide when writing up the results from a cognitive interviewing study. This function of Q-Notes also fills in data information entered into a Q-Notes project. For example, it can generate a table that summarizes the demographic characteristics of the study's sample of respondents. It can also generate a template for question-by-question review that includes a frequency table of selected response options along with the thematic schema for each question. Example 8.10 shows what the question-by-question review looks like for one of the vision questions. This printout provides the data necessary to begin writing the results of the cognitive interview project. The researcher can then copy the printout into a word processor application and begin writing conclusions about that question and how it performed.

8.3 PROJECT MANAGEMENT FEATURES

This chapter has, so far, shown how Q-Notes supports the five levels of analysis that make up a cognitive interviewing project. However, some of the set-up and additional features were not specifically mentioned. One advantage Q-Notes has over other qualitative software packages is that Q-Notes was designed specifically for conducting cognitive interviewing projects. The organization and layout of projects is consistent across projects. Q-Notes provides an intuitive process for setting up new projects.

EXAMPLE 8.10 WRITE REPORT: QUESTION-BY-QUESTION REVIEW

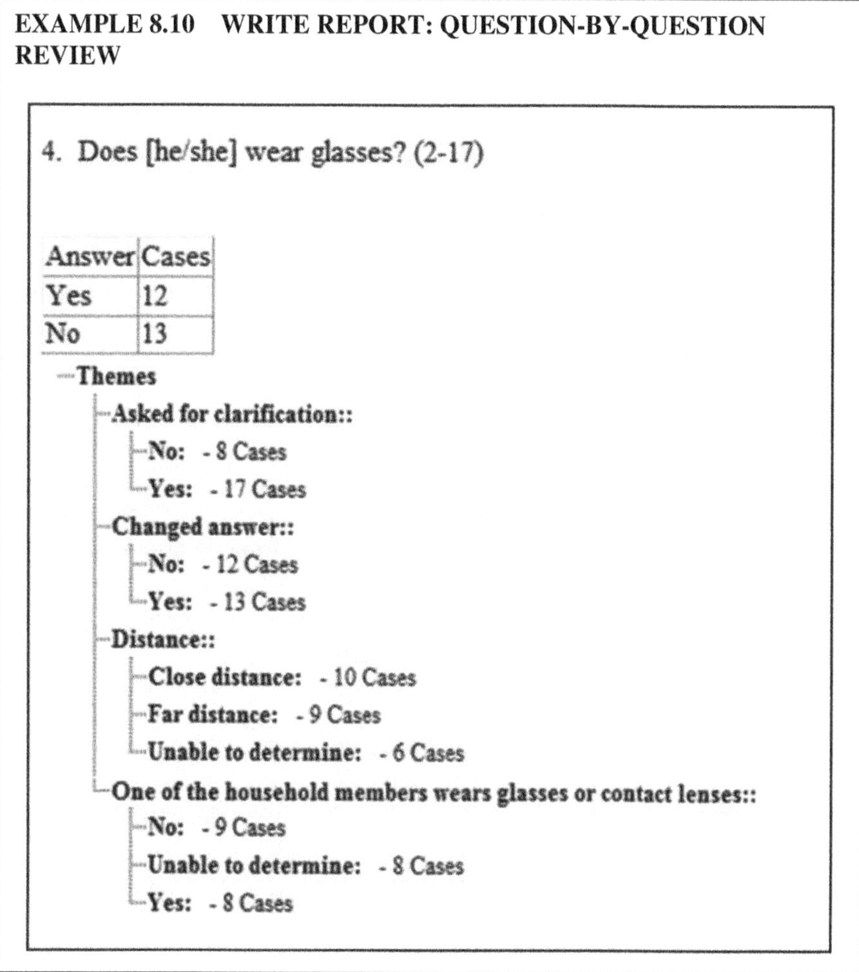

4. Does [he/she] wear glasses? (2-17)

Answer	Cases
Yes	12
No	13

Themes
- **Asked for clarification::**
  - **No:** - 8 Cases
  - **Yes:** - 17 Cases
- **Changed answer::**
  - **No:** - 12 Cases
  - **Yes:** - 13 Cases
- **Distance::**
  - **Close distance:** - 10 Cases
  - **Far distance:** - 9 Cases
  - **Unable to determine:** - 6 Cases
- **One of the household members wears glasses or contact lenses::**
  - **No:** - 9 Cases
  - **Unable to determine:** - 8 Cases
  - **Yes:** - 8 Cases

8.3.1 Streamlined Communication

When setting up a project, the project manager can enter general information. This information should provide guidance, instructions, or contact information for interviewers on a project. This information is displayed on the project home screen, so it can be viewed by users every time they enter the project. Example 8.11 displays the project information summary, which is displayed every time a user logs into a project within Q-Notes. This ensures that any updates or changes are brought to the attention of all users on the project.

8.3.2 Interview Data Collection

Q-Notes provides a standard set of demographic fields that can be used to describe respondents or interview settings. These fields are displayed for each interview and

EXAMPLE 8.11 PROJECT INFORMATION SUMMARY

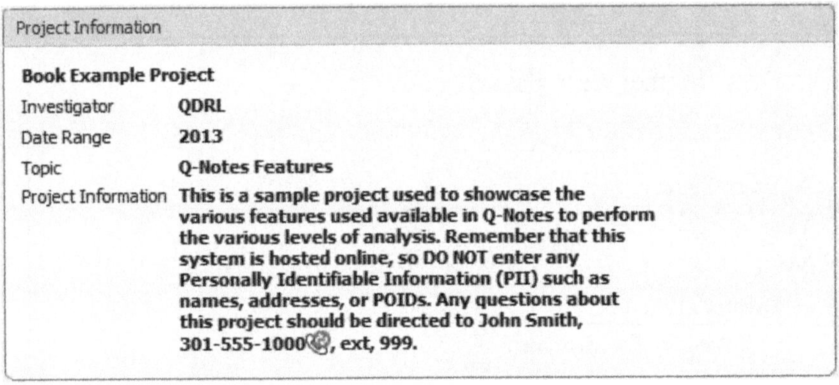

Project Information

Book Example Project

Investigator	**QDRL**
Date Range	**2013**
Topic	**Q-Notes Features**
Project Information	**This is a sample project used to showcase the various features used available in Q-Notes to perform the various levels of analysis. Remember that this system is hosted online, so DO NOT enter any Personally Identifiable Information (PII) such as names, addresses, or POIDs. Any questions about this project should be directed to John Smith, 301-555-1000, ext, 999.**

can also be used in the analysis functions. Example 8.12 depicts how Q-Notes allows a user to choose the fields that fit the project's needs. These fields can be set as optional or required for input when an interviewer is entering data.

EXAMPLE 8.12 PROJECT DEMOGRAPHIC FIELDS

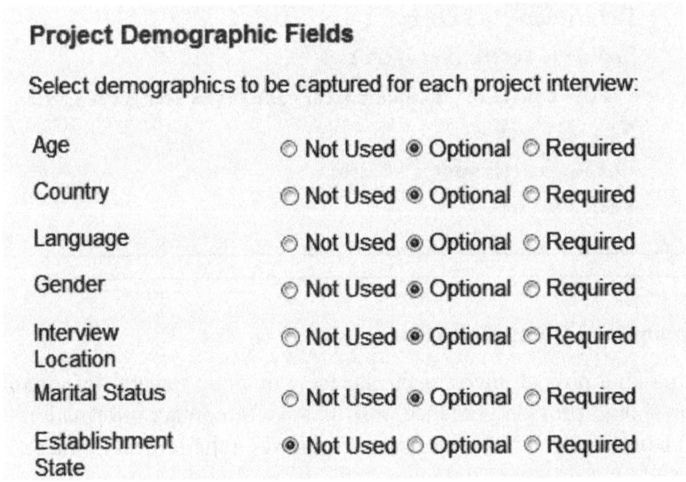

Project Demographic Fields

Select demographics to be captured for each project interview:

Age	○ Not Used	◉ Optional	○ Required
Country	○ Not Used	◉ Optional	○ Required
Language	○ Not Used	◉ Optional	○ Required
Gender	○ Not Used	◉ Optional	○ Required
Interview Location	○ Not Used	◉ Optional	○ Required
Marital Status	○ Not Used	◉ Optional	○ Required
Establishment State	◉ Not Used	○ Optional	○ Required

When adding new information related to a respondent, the selected field will be displayed as seen in Example 8.13. Q-Notes also provides a space to enter and organize all the details pertaining to an individual interview, including the date and location of the interview, as well as the person who conducted the interview. This feature helps manage and organize the data, and it documents details pertaining to the sample of respondents.

EXAMPLE 8.13 ADDING RESPONDENT DEMOGRAPHICS

Interview Details

Interview Information

RespondentID: CD14

Interview Date: 12/1/2013

Interviewer: MMassey

Interview Location: Laboratory

Gender: Male

Age: 5

Marital Status: Never Married

Country:

Language of Interview: English

Cancel Update

Interview Summary Notes:

r was a young mother with 2 children. Focus of interview is older son, 5 years old. Child has a blurry eye and is supposed to wera a patch. Surgery was done, but did not correct the problem completely. r states that she does everything for the child and he is never out of her company. Attends a school for children with functional difficulties two days a week. r also

Check Spelling

Analysis

Project Themes
- disability status
  - no
  - *yes*
- Disability Type
  - Mobility
  - Hearing
  - *Vision*
  - Learning

8.3.3 Respondent Descriptors

If the standard set of Q-Notes demographic fields is not appropriate for a particular project, Q-Notes provides the ability to add "Project Themes" or "Respondent Descriptors." These are created by the project lead and are used to identify information about the respondent or the interview as a whole. For example, in the child disability project, the field "disability type" was created with appropriate categories ("Mobility," "Hearing," "Vision," and "Learning"). All interviews could then be assigned one or more of these categories, based on a screener that was used. The summary notes in Example 8.13 mention how the child has a blurry eye and wears a patch. So the researcher selected the "Vision" category under the disability type. The categories applied here can then be used to group respondents in the analysis functions.

8.3.4 Controlled Access by Project

During project set-up, Q-Notes allows project managers to control access to the project in two different ways. First, the project manager decides who can see each project. This way, users only have access to the projects they are working on. Second, within a project, the project manager can set privilege levels or roles for what each assigned user can do. Table 8.1 lists the different user roles and the major functions within Q-Notes that the different roles can perform.

TABLE 8.1 Q-Notes User Role Types

Available Functions/Roles	*Observer*	*Interviewer*	*Analyst*	*Manager*
View all interviews and data	✓	✓	✓	✓
Add or edit their own interviews		✓	✓	✓
Apply themes to their own interviews		✓	✓	✓
Create themes			✓	✓
Code themes to others' interviews			✓	✓
Add questions				✓
Change project options				✓
Add users to project				✓

8.3.5 Adding Questions

In the initial project set-up, the interview questions must be added. Adding questions in Q-Notes is not complicated. Example 8.14 displays the fields needed to enter a question into a project.

EXAMPLE 8.14 QUESTION SET-UP

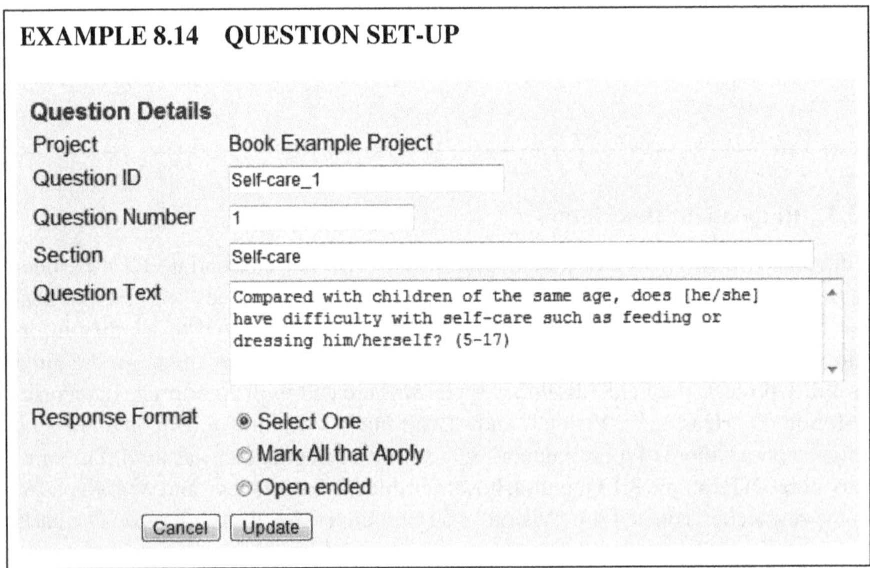

Each of the fields displayed in Example 8.14 have a specific purpose within Q-Notes. Here are the uses for each of the fields:

- **Project Question ID**—A short unique identifier for the question to be used throughout the project to refer to this question. The ID should contain a reference to the question for the interviewers to easily know what question it is. Having a project-specific prefix can also help ensure individuality for future projects.

- **Question Number—**A numeric value to set the order, or sequence, of the questions.
- **Section**—A title for current set of questions. This provides the ability to quickly view subsets of questions within the project.
- **Question Text**—The verbatim text of the question.
- **Response Format**—Refers to the type of question being asked.
  - **Select One**—Only one choice may be selected from pre-set response options.
  - **Mark All that Apply**—More than one choice may be selected from the response options.
  - **Open Ended**—Response is open ended. Allows free text to be entered as a response to the question.

After each question has been added, Q-Notes allows the project manager to define the response options, as shown in Example 8.15. When entering a new response option, Q-Notes allows the user to add a sequence, which sets the display order for the response option. Skip instructions that call attention to skip patterns for specific response options are also defined. In Example 8.15, the skip pattern for the "No difficulty" response instructs the interviewer to "Go to question 6." Skip patterns and other instructions are displayed on the Interview Guide, described earlier.

EXAMPLE 8.15 QUESTION RESPONSE OPTIONS SET-UP

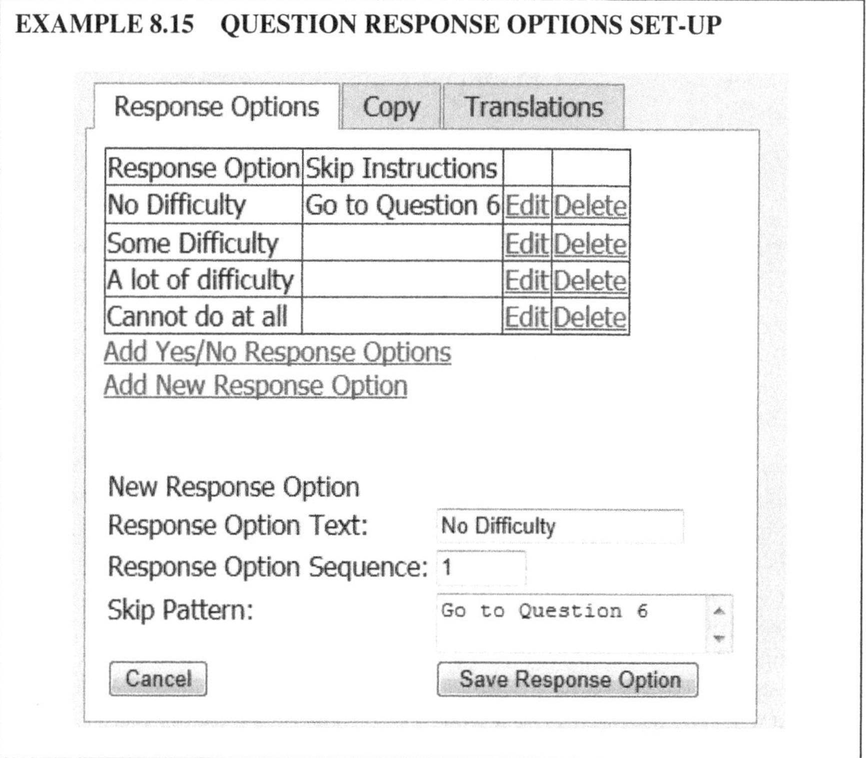

8.3.6 Question Translations

Q-Notes also simplifies the process of testing a questionnaire in multiple languages. Translated versions of questions can be entered and displayed alongside the original question. The country and language are entered along with the translated question text. Example 8.16 shows the Spanish translation (used for the interviews conducted in Belize) for the self-care question. Q-Notes captures both the country and the language for each translation. This distinction allows for researchers located in different countries, who speak the same language, to test different translations of questions, while still providing for combined analyses of the resulting data.

EXAMPLE 8.16 QUESTION TRANSLATIONS ENTRY

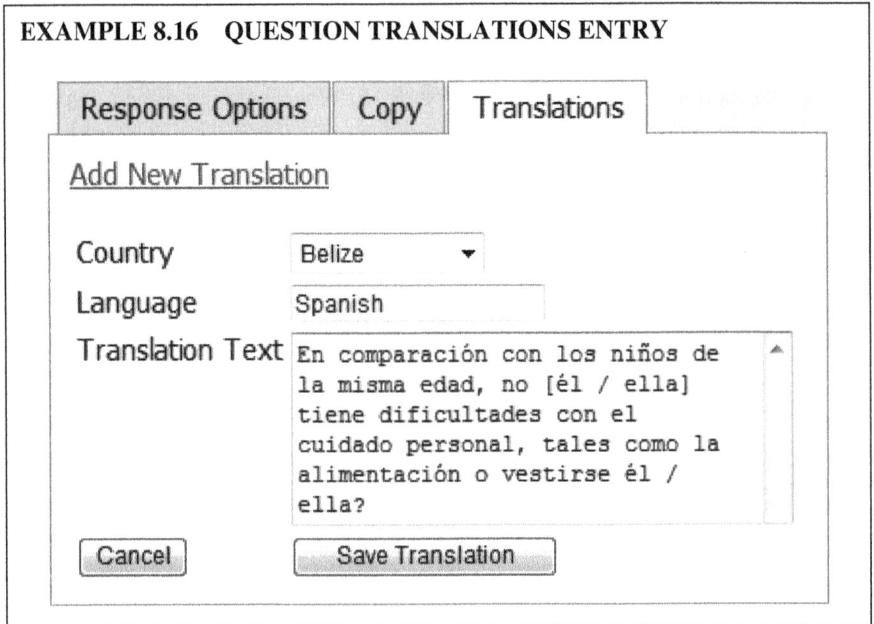

8.3.7 Coding Schemes

Q-Notes also offers the ability for behavior coding. A default list of common Interviewer and Respondent Behavior Codes is available. The list can be adjusted on a project-by-project basis. Adding the behavior codes displays a "Coding" area on pages where notes are entered. The codes can also be displayed on the Interview Guide, as seen in Example 8.17.

Q-Notes provides a centralized location to support a cognitive interviewing study and manage all the data, along with a variety of tools and functions to conduct a full and thorough analysis. Finally, at the end of the process, Q-Notes provides the basis to write a report of the findings. The knowledge gained from each project is useful to a broad audience. Eventual users of the data and other surveys can gain valuable insight from the findings about the subject at hand. Survey methodologists can also learn more about the methods used, which can lead to a better understanding of the method and improvements to the method over time. The Q-Bank application achieves this goal by making question evaluation reports public.

EXAMPLE 8.17 INTERVIEW GUIDE WITH BEHAVIOR CODES

1. Compared with children of the same age, does [he/she] have difficulty with self-care such as feeding or dressing him/herself?

Response Options	Skip Pattern
No Difficulty	Go to Question 6
Some Difficulty	
A lot of difficulty	
Cannot do at all	

Interviewer Behavior Codes

- ☐ Question read correctly
- ☐ Skip Error (for non-computerized instruments)
- ☐ Question read with minor change
- ☐ Question read with major change
- ☐ Follow-probe used incorrectly (biasing, etc.)

Respondent Behavior Codes

- ☐ Respondent Interrupts question reading
- ☐ Respondent requires Repeat)
- ☐ Respondent requests Clarification
- ☐ Respondent expresses Uncertainty
- ☐ Respondent provides Uncodeable answer

2. [If child uses a hearing aid] Does [he/she] have difficulty hearing when using his/her hearing aid(s)?

8.4 Q-BANK: MAKING COGNITIVE INTERVIEW FINDINGS PUBLICLY ACCESSIBLE

Once a research project is complete and the report has been written, the final task is to disseminate the findings. Developed by The National Center for Health Statistics, Q-Bank is a resource for disseminating the findings of question evaluation projects. Q-Bank provides access to evaluated survey questions and reports. Q-Bank houses question evaluation reports from a number of statistical agencies and research organizations. It links each question to its test findings and the original evaluation report. Q-Bank's mission is to improve the quality and usefulness of survey data through the sharing of knowledge about survey questions. Q-Bank is used by the following types of users:

- *Data Analysts* can use Q-Bank to understand the underlying concepts that questions are measuring.
- *Questionnaire Designers* can use Q-Bank to find questions that have already been evaluated on a variety of question topics, and the strengths and weaknesses of alternate versions of questions.
- *Survey Methodologists* can use Q-Bank as a way to share their findings with other survey researchers, and also learn about the strengths and weaknesses of different methods.
- *Survey Managers* can use Q-Bank to demonstrate their survey collects the data it purports to collect.

8.5 Q-BANK FEATURES

Q-Bank contains numerous question evaluation reports and even more questions. Q-Bank provides search functions to quickly direct a user to the information they need.

8.5.1 Searching for Questions

Q-Bank makes it easy to find questions on a variety of topics. All of these questions have been cognitively tested and the findings from those projects are available to the public. Users can search by keyword directly from the Q-Bank home page, as seen in Example 8.18. The quick search feature in Q-Bank also allows the user to search for the reports directly.

EXAMPLE 8.18 QUICK SEARCH EXAMPLE

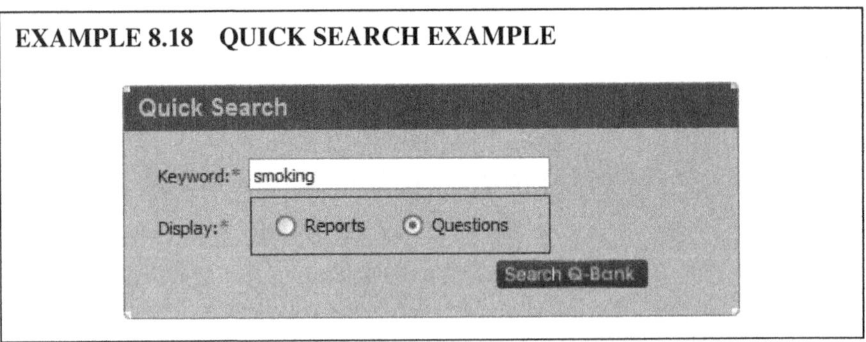

8.5.2 Advanced Search

In addition to the quick search, Q-Bank allows users to search using more criteria on the question search screen. Additional fields include:

- Question Topic and Sub-topics.
- Testing Agency (Agency which conducted the evaluation).
- Survey Title (Survey on which the question appears).
- Keywords (Words or phrases in the text of the question or the construct itself).

Example 8.19 displays a search conducted by a user. The user searched for health-related smoking questions that were tested for the National Health Interview Survey (NHIS).

EXAMPLE 8.19 ADVANCED QUESTION SEARCH EXAMPLE

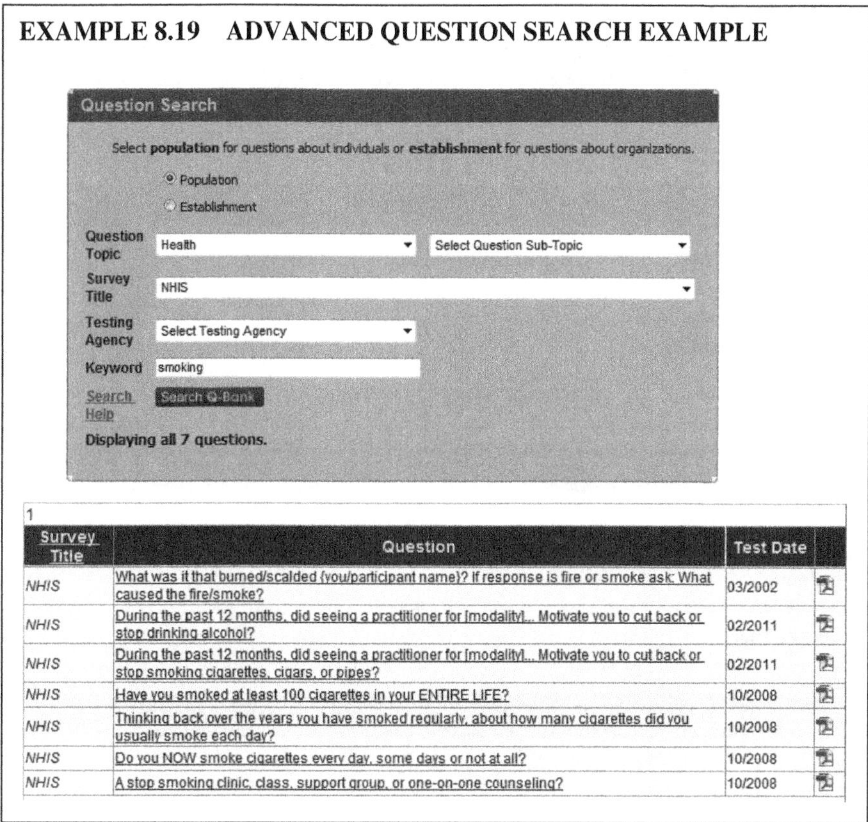

Survey Title	Question	Test Date
NHIS	What was it that burned/scalded {you/participant name}? If response is fire or smoke ask: What caused the fire/smoke?	03/2002
NHIS	During the past 12 months, did seeing a practitioner for [modality]... Motivate you to cut back or stop drinking alcohol?	02/2011
NHIS	During the past 12 months, did seeing a practitioner for [modality]... Motivate you to cut back or stop smoking cigarettes, cigars, or pipes?	02/2011
NHIS	Have you smoked at least 100 cigarettes in your ENTIRE LIFE?	10/2008
NHIS	Thinking back over the years you have smoked regularly, about how many cigarettes did you usually smoke each day?	10/2008
NHIS	Do you NOW smoke cigarettes every day, some days or not at all?	10/2008
NHIS	A stop smoking clinic, class, support group, or one-on-one counseling?	10/2008

After the search is completed, the questions that match are listed with the date that the question was tested, as seen in Example 8.19. A question is entered into Q-Bank each time that it is tested. Therefore, questions that have been tested multiple times will be repeated in the list for as many times as it was tested. This will indicate to the Q-Bank user that more than one set of test findings are available. From the search results list, a user may select one particular question which will then direct the user to the Question Details page, containing more information about the question.

8.5.3 Question Details

The Question Details page as shown in Example 8.20 displays much more information about the selected question. The page provides the question text, any introductory text, and the response categories. In addition, the page relates the type of evaluation that was conducted along with the error categories associated with the question.

EXAMPLE 8.20 QUESTION DETAILS EXAMPLE

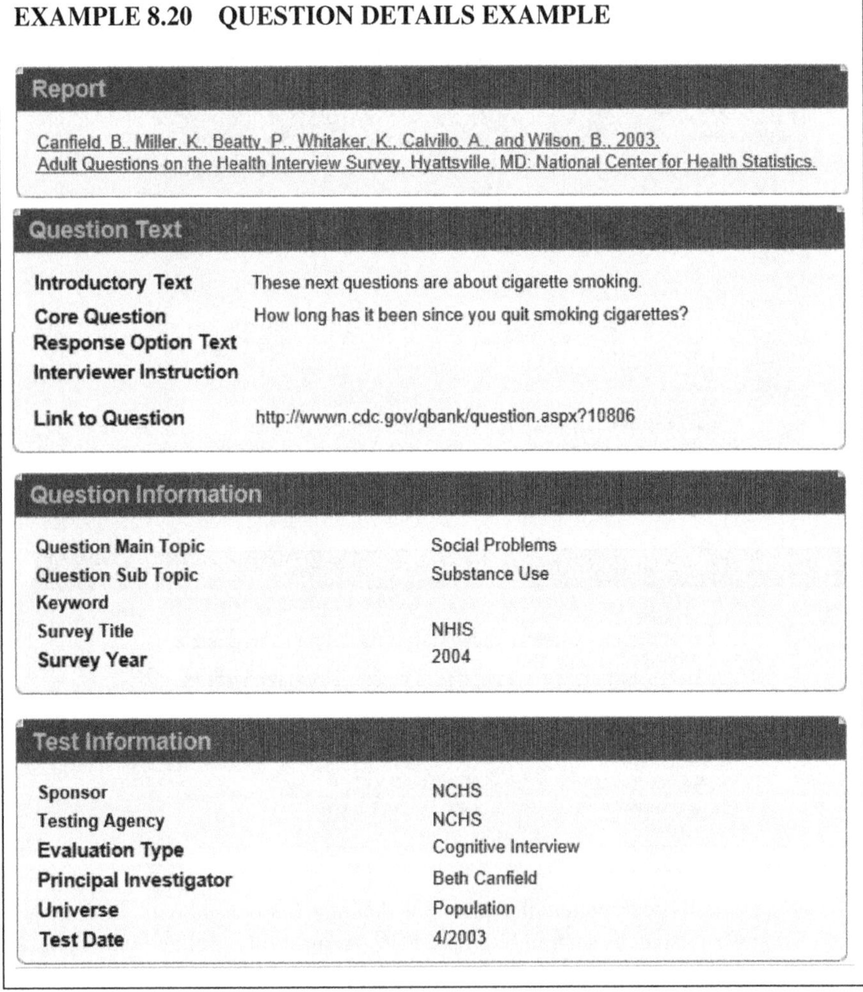

The Question Details page contains a link so that the user can refer to the final report and read, in detail, what was found as well as how the evaluation was actually conducted. Clicking the citation will open the full PDF report containing the question, and go directly to the page in the report for the question, as shown in Example 8.21.

EXAMPLE 8.21 FINAL REPORT TEXT EXAMPLE

PRT.3 Do you currently see a practitioner for [modality] more, less, or about the same as you did one year ago?

1) More
2) Less
3) About the same

Most respondents appeared not to have trouble answering this question, though one recurring issue did emerge. Specifically, for those who no longer practice the modality the options did not always appear so straightforward. Several of these respondents explained that at the time of the interview they were no longer seeing the practitioner and thought that the question might not be applicable. For instance, one respondent who was answering for his son who had seen a practitioner in the past 12 months but was not seeing a practitioner at the moment, hesitated before answering "more." He answered this way, however, because a year ago he was not seeing a practitioner at all, but he *had* seen a practitioner over the course of the previous 12 months. It may be that the "one year ago" reference is vague or unclear, or perhaps too broad. However, it is not clear what a solution to this might be.

For respondents who see a practitioner only sporadically or on an "as needed" basis, this question may also be difficult to answer. For instance, one respondent who had used

8.5.4 Value of Q-Bank

It should be emphasized that Q-Bank is not a database of ready-to-use survey questions. Q-Bank users would be remiss if they chose to use a question appearing in Q-Bank without first examining what was found in the evaluation. Questions appearing in Q-Bank are the original questions—questions that were tested—and are not the revised questions. Consequently, some of the questions in the database may not capture their intended construct.

That being said, Q-Bank serves survey researcher in a variety of ways. Specifically, it is a tool for:

1. Designers tasked with developing new questionnaires who would be assisted in seeing what did and did not work in previous questions and *why* they did not work,
2. Subject matter analysts seeking additional information about the data in order to help make sense of relationships that they have found or have not found,
3. And, survey methodologists doing research on question design and response error.

8.6 Q-BANK: CHALLENGES FOR THE PAST AND FUTURE

As a centralized, multiple-search warehouse of survey questions with links to pre-test information, Q-Bank is envisioned to:

1. Provide an invaluable resource to topic specialists and other non-design specialists tasked with developing new survey questions.
2. Promote standardization and comparability across federal surveys.
3. Facilitate the comparison of cognitive testing findings over time, in different settings, with different sub-populations and across federal agencies, thereby providing insight into the reliability and validity of cognitive testing and other methods of question evaluation.
4. Provide ability to analyze specific types of question characteristics that may contribute to response error.
5. Provide an additional resource for analysts in the interpretation of survey data.
6. Increase awareness of response error as part of total survey error, generating a research agenda that ultimately seeks to quantify response error and improve the overall methodology of survey research.

Finally, there are two prominent challenges for Q-Bank of the future. First, the development of Q-Bank as a tool must continue. To thrive, Q-Bank must grow in terms of the amount of data it houses, the number of researchers it serves, as well as the number of pre-test methodologies that it will accommodate. Second, the original goal of serving a methodological research agenda must be maintained. If only a repository of questions, it is of much less value to survey research. The value of Q-Bank lies in its potential to advance the field of question design and evaluation.

8.7 CONCLUSION

Built on principles of qualitative analysis, and designed specifically for cognitive interviewing, Q-Notes simplifies the processes of conducting a cognitive interviewing study. Q-Notes allows various researchers working on the project to enter and access the interview data in a centralized location. Analysts are then able to view data and enter thematic schema directly into the program. All researchers are then able to see how the analyst has reached conclusions. In this way, Q-Notes is not only useful for assisting with analysis but also in creating an audit trail that links raw data to study conclusions.

Q-Bank educates the public and other data users about what surveys questions are actually capturing. Without this information, data users may make faulty conclusions when interpreting survey data. Policymakers and public health officials rely on survey data to develop interventions and assist communities. As future questionnaires and surveys are written, survey designers can use Q-Bank to find previously examined versions of questions or subject areas, allowing them to understand the strengths and weaknesses of specific questions and to determine which questions are most suitable to their requirements.

Q-Notes can be accessed at http://wwwn.cdc.gov/qnotes

Q-Bank can be accessed at http://wwwn.cdc.gov/qbank

9 Cognitive Interviewing in Mixed Research

ISABEL BENITEZ BAENA and JOSÉ-LUIS PADILLA
University of Granada, Spain

9.1 INTRODUCTION

Cognitive interviewing methodology, with its ability to reveal the substantive meaning behind a survey statistic, can provide critical insight into question performance. Nonetheless, cognitive interviewing methodology has its own limitations and cannot provide an entire picture of question performance. While cognitive interviewing can show that a particular interpretive pattern does indeed exist, it cannot determine the extent or magnitude to which that pattern would occur in a survey sample. Nor can cognitive interviewing studies reveal the extent to which variation of interpretive patterns would occur across various groups of respondents. In addition, the method cannot fully determine the extent to which respondents experience difficulty when attempting to answer a question. In short, as a qualitative methodology, cognitive interviewing studies lack the ability to provide quantitative assessment—a component particularly essential to the field of survey methodology. Integrating quantitative methods into cognitive interviewing studies can work to resolve these limitations.

This chapter illustrates how cognitive interviewing studies, as presented in this book, might be combined with quantitative methodologies in order to reveal a more detailed picture of question performance that neither methodology could portray on its own. The chapter first describes the history and key components of mixed research. It then describes how mixed research can inform question evaluation, particularly as it applies to cognitive interviewing. Finally, the chapter presents three examples of mixed research involving cognitive interviewing. The examples illustrate how mixed research can provide a broader understanding of question performance, thereby allowing for better decisions regarding question design and data quality—that is, whether a question should be altered or if the resulting variable should be utilized or interpreted in a particular way.

Cognitive Interviewing Methodology, First Edition.
Edited by Kristen Miller, Stephanie Willson, Valerie Chepp, and José-Luis Padilla.

9.2 THE MIXED RESEARCH PARADIGM: CHARACTERISTICS AND DESIGN

Tashakkori and Teddlie (1998) were the first to define mixed research, presenting it as research that combines quantitative and qualitative methods in order to obtain more sophisticated findings for complex research problems. The paradigm sought to reconcile the traditional debate between quantitative and qualitative research by confronting limitations of methodological exclusivity (Reichardt and Rallis 1994). As the field developed, various ideas and debates emerged, introducing new—and sometimes conflicting—conceptualizations. Through debate, however, consensus centered upon a significant theme: research is "mixed" when the differing methodologies are utilized in a complimentary way in order to achieve the research objective. This theme gave way to arguably the most agreed upon definition posed by Tashakkori and Creswell (2007): the investigator collects and analyzes data, integrates the findings, and draws inferences using both qualitative and quantitative approaches. With the advancement of the mixed methods field, researchers have established specific principles and concepts that further define the field (Small 2011). The mixed research vocabulary and main principles were generated in order to standardize the way studies are performed and reported.

The concepts of pragmatism and integration have become particularly significant concepts within the field of mixed method research, helping to specify the parameters of the methodology further. *Pragmatism* refers to the prioritization of the research objective over the training of the researcher—whether it be quantitative or qualitative (Johnson and Onwuegbuzie 2004; Johnson and Christensen 2008). The objective of the study must be clearly defined, and the sub-objectives that each method will meet must be articulated; this pragmatic focus, in turn, justifies the choice of methods. *Integration* refers to the requirement of fully combining qualitative and quantitative methods toward a cohesive understanding of a phenomenon. It goes beyond the separate revelations of a qualitative and a quantitative understanding and rather requires that the two methodologies are used cohesively to provide a broader and more detailed understanding of the research objective.

As various qualitative and quantitative methods are utilized within a mixed research study, numerous types of study designs can be used (Creswell 1995; Greene et al. 1989; Tashakkori and Teddlie 1998). One of the most widely used design classifications is Creswell's (1995), which includes two dimensions: sequentiality and dominance. In regard to the dimension of *sequentiality*, studies can either be simultaneous or sequential, that is, they can either apply the two methods concurrently or in stages. *Dominance* refers to priority, that is, one method may be more dominant than the other, or both methods can be equally relevant. As Table 9.1 shows, six types of basic designs are obtained under Creswell's (1995) classification schema, which become nine possible designs when order is considered.

Creswell (1995) presented this classification schema using different combinations of symbols and punctuation. Capital letters are used to point out the dominant design element and lower case letters are used for less dominant components. The

TABLE 9.1 Possible Designs when Sequentiality and Dominance are Combined

Basic designs		Order influence
Simultaneous	Sequential designs	(Sequential studies)
QUAN/QUAL	QUAN+QUAL	QUAL+QUAN
QUAN/qual	QUAN+qual	Qual+QUAN
Quan/QUAL	Quan+QUAL	QUAL+Quan

plus symbol indicates that the components are sequential and the slash indicates simultaneous phases.

9.2.1 Cognitive Interviewing Studies and Research Design

Cognitive interviewing studies can take place within a variety of research designs, including Monomethod designs that are not "mixed" because they are combined with other qualitative methods. Figure 9.1 illustrates these different designs.

Monomethod studies. Cognitive interviews are conducted in order to obtain information about life experiences of respondents that account for the various constructs captured by a particular survey question. As part of a survey question validation project, cognitive interviewing data could be complemented with data from other pre-test methods. If evidence comes from other qualitative pre-test methods, the question evaluation project is a "Monomethod" research study, since only qualitative methods are used in the research design. For example, Cortes et al. (2007) applied different qualitative procedures to develop a culturally and linguistically appropriate version of a mental health outcome measure by using focus groups and cognitive interviews with different types of Spanish speakers. The integration of results from cognitive interviews and focus

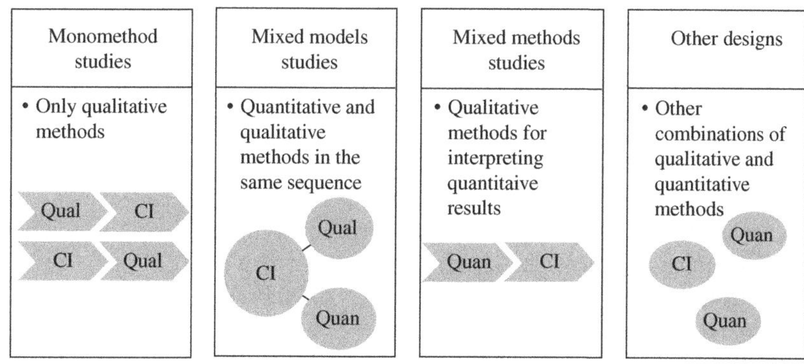

FIGURE 9.1 Research designs

groups was useful for identifying commonly used vocabulary and developing culturally appropriate measures. While Monomethod studies combine different methodologies in a complimentary way, they are not mixed research since they do not integrate quantitative and qualitative methodologies. However, it can still be useful to combine different qualitative methods in order to achieve a research objective.

Mixed model studies. Cognitive interviewing data are sometimes analyzed following quantitative approaches, especially when researchers try to identify problems or difficulties experienced by respondents. However, quantitative results obtained from qualitative data collected via a purposive sample are not generalizable and therefore may not be appropriate or advantageous. When these analytic approaches are implemented, interpretation patterns are categorized and themes and sub-themes counted. From a mixed research perspective, this kind of study can be seen as a "Mixed Model" study in which the same set of data is analyzed using qualitative and quantitative analytic strategies. For example, Blanden and Rohr (2009) investigated patient satisfaction by analyzing cognitive interviewing data as both open-ended and close-ended responses.

Mixed methods studies. Finally, qualitative results from cognitive interviews can be used for interpreting statistical results. In these studies, a mixed method design allows researchers to connect different types of data to achieve integrated results. Mixed methods studies are the most adequate design when there are two types of data that can be combined in a complementary way in order to reach a shared conclusion.

9.3 MIXED METHOD RESEARCH AND SURVEY QUESTION EVALUATION

While mixed method research has expanded into numerous fields within the social sciences, it remains relatively absent in the field of survey methodology. A mixed research design that combines a qualitative question evaluation method such as cognitive interviewing with a quantitative method, such as latent class analysis or item response theory (IRT), can be highly advantageous (Presser et al. 2004). Cognitive interviewing studies determine the patterns of interpretation, that is, the construct(s) captured within a statistic, while quantitative methods can ascribe a quantity to those patterns. In short, the qualitative component answers questions regarding what and why, while the quantitative component can answer how much. In an edited volume of question evaluation methods, Reeve (2011) states that an IRT model may require a qualitative component to "identify why a particular item or set of items perform in a way inconsistent with what was expected" (p. 122). Further, question evaluation models that rely on the assumption of latency should utilize the method of cognitive interviewing to confirm that the assumed construct is indeed a correct assumption. In this same volume, editors Madans et al. (2011) conclude that, "Since each evaluation method addresses a different aspect of quality, the methods should be used together ...

important synergies can be obtained if evaluations are planned to include more than one method and if each method builds on the strengths of the others" (p. 2).

Mixed research is particularly beneficial when attempting to understand Total Survey Error. A central concept in the field of survey methodology, total survey error aims to understand all major sources of error in a survey. This refers to the accumulation of all errors that may arise in the design, collection, processing, and analysis of survey data (Biemer 2011). The sources of survey errors are numerous and include deficiencies in the sampling frame, the sampling process, the interview, the interviewers, respondents, and coding. The total survey error paradigm aims to develop a metric—the mean square error—which allows not only quantifying the total error and the contribution of each of the sources, but also optimizes the survey design by allocating the survey resources efficiently in order to minimize the error of at least the most important estimates. However, a major difficulty for achieving that goal is how to combine the statistical and non-statistical components of errors (Groves et al. 2009).

The form of error relevant to question evaluation is specification error, which addresses the relationship between the construct being measured and the question. Biemer (2011), for example, states that specification error occurs when the concept implied by the question differs from the concept that should have been measured. This difference leads to measuring the construct wrongly and estimating the parameter incorrectly, thereby undermining the validity of the inference. Biemer (2011) locates the cause of the specification error in poor communication between the researcher and the questionnaire designer. A mixed method approach that includes cognitive interviewing can not only define and measure the degree of miscommunication, but also allow for the integration of statistical and non-statistical classifications of errors. Groves and Lyberg (2011) also move in this direction by pointing out one of the limitations of the total survey error framework, that is, the exclusion of key concepts of quality such as validity and feasibility. As a task for the future, they call for more communication between scientists studying the causes of errors and those who statistically model them.

Applying cognitive interviews in combination with other pre-test methods within a mixed research framework moves in this direction. The following three cases illustrate how cognitive interviewing can be paired with three different quantitative methods: survey field testing, differential item functioning (DIF) analysis, and psychometric scaling.

9.3.1 Case 1: Cognitive Interviewing and Survey Field Testing

Combining cognitive interviewing and survey data in a QUAL+QUAN design (Table 9.1), Miller and Maitland (2010) show how a mixed method study can provide evidence of the extent to which findings from cognitive interviews are present in a survey sample, and if different linguistic versions of questions in cross-cultural surveys are capturing real differences. The authors focused on questions intended to measure anxiety for an international collaborative research project in six Asian countries.

This study was part of a larger project sponsored by The Washington Group and the United Nations Economic and Social Commission for Asia and the Pacific Statistics Division (WG/UNESCAP) to develop internationally comparable disability questions. The project involved a mixed method design utilizing a two-step approach for question evaluation: a cognitive interviewing study and a field test. First, 143 total semi-structured, qualitative cognitive interviews were conducted in six participating countries (Cambodia, Kazakhstan, Maldives, Mongolia, the Philippines, and Sri Lanka) in order to understand the ways in which each question performed. Based on the analysis of the cognitive interviews, follow-up probe questions were developed and placed on the field test questionnaire. Each country then conducted approximately 1000 standardized survey interviews drawn from a random sample. Resulting survey data from the follow-up probe questions were used to examine the extent of valid and non-valid interpretive themes in a cross-national environment.

9.3.1.1 Cognitive Interviewing Results Findings from the cognitive interviewing component established hypotheses to be examined in the second component, the field test. While cognitive interviews uncovered the specific patterns of interpretation, the field test was used to understand the extent to which those patterns existed. The set of anxiety questions included in the cognitive interviews are presented in Figure 9.2. As a set, the intent of the questions is to place respondents along a severity continuum comprised of various dimensions of anxiety (i.e., frequency, intensity, and consistency).

In considering the constructs captured by the anxiety questions, respondents considered a range of feelings and experiences that they recognized as anxiety—or rather, what they believed the question was asking in terms of feeling worried, nervous, or anxious. For the most part, the feelings and experiences considered by respondents can be seen as various aspects of the intended concept of anxiety, though ranging in severity. These aspects include:

1. clinical anxiety, whereby respondents described being diagnosed by a medical professional,
2. elements of depression, whereby respondents spoke about being overly sad, wanting to stay in bed, or being unable to perform daily activities, and

Anxiety

1. How often do you feel worried, nervous, or anxious? Daily, weekly, monthly, a few times a year, or never?
2. Do you take medication for anxiety?
3. Thinking about the last time you felt anxious, how would you describe the level of anxiety? Mild, moderate, or severe?
4. Thinking about the last time you felt anxious, was the anxiety worse than usual, better than usual, or about the same as usual?

FIGURE 9.2 Anxiety questions examined in the cognitive interviewing study

3. stress-related worry, which respondents connected to work (e.g., heavy workloads, deadlines, and performances), family or relationship problems, crime, or concerns about their economic future and physical well-being.

A fourth and problematic theme, however, was that some respondents spoke about their anxiety as being a positive characteristic. These respondents, it appeared, interpreted the question as asking about being excited, energetic, or looking forward to the future.

Based on these findings, a few changes were made to the questions. Most importantly, probe questions were placed after two revised anxiety questions so that it was possible to understand the nature of respondents' reported anxiety. Responses to probe questions would reveal the extent of out-of-scope interpretations. In addition, the probe questions would reveal whether the questions operated similarly across the six countries. Finally, the probe questions would allow investigators to better understand how the two questions—frequency and intensity—work together in placing respondents on a severity continuum.

The revised frequency and intensity questions along with the probe questions are presented in Figure 9.3.

9.3.1.2 Field Test Results Respondents first reported how frequently they feel worried, nervous, or anxious. Overall, nearly half of the respondents (47.3%) in the

---

FREQUENCY How often do you feel worried, nervous or anxious?

1. Daily
2. Weekly
3. Monthly
4. A few times a year
5. Never

[*If R reports anxiety*]

INTENSITY Thinking about the last time you felt worried, nervous or anxious, how would you describe the level of these feelings?

1. A little
2. A lot
3. Somewhere in between a little and a lot

PROBES Please tell me which of the following statements, if any, describe your feelings. [Yes/No]

A. Sometimes the feelings can be so intense that my chest hurts and I have trouble breathing.
B. These are positive feelings that help me to accomplish goals and be productive.
C. The feelings sometimes interfere with my life, and I wish that I did not have them.
D. I have been told by a medical professional that I have anxiety.

FIGURE 9.3 Anxiety field test questions

field test reported that they never experienced these feelings. One in four reported experiencing the feelings a few times a year, while one in ten reported having them monthly. Nearly one in five (19.1%) respondents reported that they feel worried, nervous, or anxious either weekly or daily.

Table 9.2 demonstrates that the frequency reported varied significantly by country. For example, almost one-third (30.9%) of respondents in Kazakhstan reported that they experience the feelings weekly or daily. Similarly, one in four (26.0%) respondents from Mongolia reported the feelings at least weekly. At the other end of the spectrum, only about 10% of respondents from Sri Lanka and the Philippines reported the feelings weekly or daily. In fact, 78.4% of respondents from Sri Lanka reported that they never experienced worry, nervousness, or anxiousness.

The intensity of anxiety reported by country is shown in Table 9.3. Overall, almost one in five (19.2%) respondents reported that they experienced a lot of anxiety the last time they had these feelings. The intensity of anxiety reported also varied significantly by country.

Of those not reporting anxiety at least a few times a year, one-third (34.8%) of respondents from Sri Lanka and 40.9% of respondents from Maldives described the level of these feelings as "a lot." The level of these feelings is much lower in the other countries. No more than 16.9% in any of the remaining countries described the level of their feelings as "a lot." And only 7.2% in the Philippines described the level of these feelings as "a lot."

The field test data, as presented in these tables, suggests that Sri Lanka has the least amount of anxiety. Although, when Sri Lanka respondents do report anxiety, they are likely to report high intensity. The question, then, is how much do these data reflect the "true" anxiety levels of each country? How comparable are these measures? Do the people of Sri Lanka actually have fewer cases of anxiety? Estimates from these two questions alone cannot determine comparability. Analysis of the probe questions helps to clarify these issues.

Table 9.4 presents the reporting of probe questions by country. The table indeed reveals considerable variation across countries, particularly those describing their anxiety as positive.

Eight in ten (82.5%) respondents from Mongolia reported that their feelings are positive and help them to accomplish goals or be productive. In contrast, only 12.5% of respondents from Sri Lanka report the same.

Seven in ten (72.3%) respondents from Cambodia reported that the feelings can be so intense that their chest hurts or they have trouble breathing, whereas only one in five respondents from Kazakhstan and the Philippines reported the same. Roughly two-thirds or more of respondents from Cambodia, Sri Lanka, and Mongolia reported that the feelings interfere with their life. Relatively few respondents reported any kind of clinical diagnosis for their anxiety. This is especially true in Sri Lanka where only 3% reported a diagnosis.

Although these data alone do not produce a complete picture of the questions' performance, it does indicate that the questions perform somewhat differently across countries as well as the way in which they perform differently. This analysis helps to

TABLE 9.2 Frequency of Anxiety by Country

Frequency (%)	Kazakhstan	Cambodia	Sri Lanka	Maldives	Mongolia	Philippines	All countries
Never	31.9	39.7	78.4	46.6	35.4	54.3	47.3
Few times a year	22.9	28.7	7.5	27.3	25.6	25.3	23.0
Monthly	13.6	14.5	2.7	5.7	12.8	10.8	10.1
Weekly	17.6	9.2	2.1	9.0	12.4	7.7	9.7
Daily	13.3	7.4	8.1	11.0	13.6	1.7	9.3
Refused	0.2	0.0	0.1	0.1	0.1	0.0	0.1
Do not know	**0.5**	**0.5**	**1.1**	**0.4**	**0.3**	**0.2**	**0.5**
	100	100	100	100	100	100	100
n	(1000)	(1008)	(1000)	(1013)	(1222)	(1066)	(6309)

Chi-square = 817.34, 20 df, $p < .05$

TABLE 9.3 Intensity of Anxiety by Country

%	Kazakhstan	Cambodia	Sri Lanka	Maldives	Mongolia	Philippines	All countries
A little	64.0	62.7	54.4	39.4	65.6	83.5	62.4
Closer to a little	4.9	2.0	2.0	1.7	4.3	1.4	3.0
In between	10.2	18.6	6.4	15.1	6.8	6.8	11.0
Closer to a lot	6.7	2.2	2.0	1.1	5.1	0.4	3.3
A lot	12.9	14.1	34.8	40.9	16.9	7.2	19.2
Refused	0.2	0.0	0.0	0.0	0.3	0.0	0.1
Do not know	**1.2**	**0.5**	**0.5**	**1.9**	**1.0**	**0.6**	**1.0**
	100	100	100	100	100	100	100
n	(675)	(603)	(204)	(536)	(785)	(486)	(3289)

Chi-square = 443.68, 20 df, $p < .05$

TABLE 9.4 Percent Reporting Various Descriptions of Anxiety by Country

Description of feelings (%)	Kazakhstan	Cambodia	Sri Lanka	Maldives	Mongolia	Philippines	All countries
Positive	50.3	47.8	12.6	51.7	82.5	32.4	53.0
Chest hurts	21.4	72.3	30.9	37.0	50.6	20.3	40.6
Interfere	52.2	65.0	85.4	54.8	72.8	33.5	59.1
Clinical	11.8	16.8	3.0	28.4	18.6	11.5	16.5

Note: Chi-square $p < 0.05$ for all rows in the table.

TABLE 9.5 Joint Distribution of Anxiety Frequency and Intensity

	A few times a year	Monthly	Weekly	Daily	DK/REF
A little	1087	423	328	214	1
In between	179	126	161	104	0
A lot	163	86	122	259	0
DK/REF	22	3	3	7	1

make sense of and interpret the survey data for a more realistic portrayal of anxiety in these six different countries.

The next step in analyzing the field data was to examine the probes within categories of the anxiety questions. As was learned through analysis of the cognitive interviews, respondents appeared to experience and relate their feelings of anxiousness both in terms of frequency and intensity. Table 9.5 shows the joint distribution of the anxiety frequency and intensity questions. Intuitively, the seriousness of anxiety would be lowest in the upper left corner of the table and increase as one moves toward the lower right corner of the table. In addition, the table illustrates that intensity of anxiety increases with frequency. A composite of the two variables provided a multi-dimensional continuum for depicting the severity of anxiety.

Understanding the ways in which respondents characterized their feelings within each of the cells in Table 9.5 provides an even clearer picture of this relationship. Table 9.6 illustrates how the characterizations of anxiety were associated with being located in each cell in the joint distribution of frequency and intensity. The negative associations are shown in italicized text, and the positive associations are shown in non-italicized text.

Several observations can be made from this table. First, the upper left corner of the table shows that cases of anxiety involving chest pains, interference with life and clinical diagnoses decrease the likelihood of selecting the lowest levels of frequency and intensity. In contrast, anxiety described as being related to chest pains,

TABLE 9.6 Significant Relationships with Respondent Location Based on Bivariate Logistic Regression Models in each Cell (Models run for Cases NOT taking Medication)

	Frequency			
Intensity	A few times a year	Monthly	Weekly	Daily
A little	*Chest hurts*** *Interfere*** *Clinical***	*Clinical**	*Chest hurts*** *Interfere***	Interfere**
In between		*Positive**	Chest hurts**	Interfere**
A lot	Chest hurts** Interfere* Clinical**	*Positive** Chest hurts** Interfere* Clinical**	Chest hurts** Interfere** Clinical**	*Positive*** Chest hurts** Interfere** Clinical**

*$p < 0.05$, **$p < 0.005$

interference with daily life, and clinical diagnoses generally increase the likelihood of responding at the higher levels of the frequency and intensity variables. Moreover, these variables were the most associated with the highest level of the frequency and intensity variables. Finally, anxiety described as a positive attribute decrease the likelihood of selecting the highest levels of anxiety.

This analysis, then, suggests that the two variables work well together—which also confirms the finding of the cognitive interviewing study determining that frequency and intensity were salient to people's experiences with anxiety.

9.3.2 Case 2: Cognitive Interviewing and Differential Item Functioning (DIF)

Cross-lingual surveys and testing have become an important topic in survey and psychological research. Projects like the Programme for the International Assessment of Adult Competencies (PIAAC) (Organization for Economic Co-operation and Development—OECD 2011) and the Program for International Student Assessment (PISA) (OECD 2009), are just two examples of international programs that regularly evaluate and compare people around the world.

DIF analyses are conducted to identify items or survey questions that function differentially across groups of respondents, so that these items can be inspected to determine whether the observed difference is due to some real difference between the groups of respondents, or whether it may be due to some form of construct-irrelevant variance (i.e., bias). DIF occurs when respondents with the same level in the intended construct (i.e., ability, achievement, or attitude) have different probabilities of giving a particular response to an item or survey question depending on the group (e.g., country) to which they belong (Millsap and Everson 1993). Survey researchers have become increasingly interested in DIF due to the frequent inclusion of psychological scales in health and social surveys. DIF can undermine validity of survey estimates in cross-cultural or lingual surveys.

Although DIF is identifiable in the statistical sense, the causes of DIF can be evasive. Benitez and Padilla (2014) illustrate how cognitive interviewing can help to understand the causes of DIF. The combination of cognitive interviewing findings and DIF statistics within a mixed research framework allowed researchers to link variations of interpretative patterns across different linguistic groups to DIF statistical results.

The authors combined qualitative and quantitative data in a mixed method study to research attitudinal items included in the 2006 PISA Student Questionnaire (OECD 2006), comparing the US English version of the items with the Spanish version. The study included two phases: a quantitative phase in which statistical techniques were applied in order to detect DIF items and a qualitative phase intended to identify DIF causes by cognitive interviewing.

Following Creswell (1995), Benitez and Padilla (2014) applied a QUAN+QUAL design. The nomenclature reflects two sequential phases. The first phase is quantitative, but both phases are equally relevant to the study objective. In the quantitative phase the research question was: "Which items of the 2006 PISA Student Questionnaire show DIF?", and in the qualitative phase: "What do US and Spanish respondents consider when answering the items with DIF?" By combining both phases the research question for the mixed method study was: "Are the different linguistic versions of the PISA items capturing different interpretations, schooling experiences, etc., which could explain DIF effects between US and Spanish respondents?"

Two statistical techniques were selected for identifying DIF items when comparing responses to the 2006 PISA Student Questionnaire from the US and Spanish respondents: differential step functioning (DSF) and ordinal logistic regression (OLR). For DIF analysis, data were obtained from the PISA database (OECD 2006), in which responses of 17,405 participants from Spain and 4902 participants from the United States were coded. DIF analyses were done using country as the group variable. For the application of cognitive interviews, 44 participants were recruited, 24 from Spain (15 women and 9 men) and 20 of the United States (11 women and 9 men). These respondents were chosen to mimic the characteristics of participants in the PISA study—students between 15 and 16 years who were in the final stages of compulsory education.

Seven scales were selected for analysis from the 2006 PISA Student Questionnaire (OECD 2006). The items inquired about students' general and personal value of science, their interest and enjoyment of science, their self-concept of their own scientific abilities, and whether they are motivated to use science in the future. All were four-point Likert-type item scales intended to measure science related attitudes. For example, the US English item 1 of the "Enjoyment of science" scale asked "*I generally have fun when I am learning broad science topics;* strongly agree, agree, disagree, strongly disagree."

9.3.2.1 DIF Results DIF analyses were computed for all 38 items across the seven selected scales of the 2006 PISA Student Questionnaire. DIF analysis was performed in two phases, first, at item level, and then, at response-category level for those items flagged with DIF at the item level. In phase 1, polytomous DIF was analyzed using the odds ratio approach implemented in Penfield's *DSF* framework (Penfield 2005), and OLR (Miller and Spray 1993). The odds ratio Mantel–Haenszel (MH) approach to test the null hypothesis of no DSF implemented in DSF framework involves comparing the odds of successfully being in the *j*th response category or higher on an item across examinees in different groups who are matched on the construct being measured. On the other hand, OLR compares the likelihood ratio statistic estimated in the absence of DIF, the null model with total score only is compared with that obtained when the model is adjusted for the presence of DIF, the full model with group, total score, and the interaction (Zumbo 2009). In addition, an effect size classification was done by following the criteria proposed by Penfield, Alvarez and Lee (2009), and the Educational Testing Service (ETS) criterion for interpreting the DELTA index (Zieky 1993). In this case, DIF was considered "medium" when values ranged between 1 and 1.5, and "large" when values were higher than 1.5.

After data were analyzed separately for both methods (odds ratio MH and OLR), convergence across them was evaluated. Only items flagged with medium or large DIF at the item level for both statistical methods were considered to reduce uncertainty of the DIF results. Table 9.7 shows the items flagged with DIF by two methods

TABLE 9.7 Convergence across DIF Methods at Item-Level

		Ordinal Logistic Regression	
		Large	Medium
Odd-ratio MH	Large	Enj5, Eff8, Gen1, Gen3, Gen5, Per5, Ins1, Con4	Enj3, Eff1, Eff5
	Medium	Per4, Ins4, Con3	

Enj = Enjoyment of Science; Gen = General Value of Science; Per = Personal Value of Science; Ins = Instrumental Motivation to learn Science; Eff = Science Self-Efficacy; Con = Science Self-Concept.

showing medium or large DIF in the same magnitude. Items are identified by the scale abbreviations and the item number in the scale.

In phase two, DIF analysis was conducted at response-category level using only the DSF framework because DIF statistical analysis of response categories is not possible with OLR. DSF is intended to detect DIF within each step underlying the polytomous response variable.

DSF analysis at the response-category level was applied only for items flagged with large DIF at the item level. Cumulative categories with three steps were applied since attitudinal items in the Student Questionnaire of PISA 2006 are four-point Likert item scales. DSF assumes a graded response model that uses a cumulative form, because in this model the step function describes the probability that an examinee successfully advances to a score level equal to or greater than. Table 9.8 presents the DIF patterns at the response-category level for the eight items classified as showing large DIF.

Table 9.8 shows the item code of the eight items flagged with large DIF, for instance, "Gen1" stands for item 1 of the scale "General value of science." Then, DSF analysis identifies the steps in which DIF is located, the magnitude of DIF in each step, and the country group which needed more ability to cross the specific

TABLE 9.8 Summary of Differential Step Functioning Statistics

Items	Steps	Magnitude	Sign	Form	Revision
Enj5	1	Medium	US	Pervasive–convergent	Item level
	2,3	Large			
Eff8	1,2,3	Large	US	Pervasive–constant	Item level
Gen1	1,2,3	Large	US	Pervasive–constant	Item level
Gen3	1	Medium	Spain	Pervasive–divergent	Item level
	2,3	Large	US		
Gen5	1,2	Large	US	Pervasive–convergent	Item level
	3	Medium			
Per5	3	Large	Spain	Non-pervasive–constant	Score level
Ins1	2	Medium	Spain	Non-pervasive–constant	Score level
Con4	1,2	Large	US	Pervasive–convergent	Score level
	3	Medium			

step. The combination of these three dimensions allowed researchers to determine the DSF pattern. For example, item stem for Gen1 is "*Advances in broad science and technology usually improve people's living conditions. Strongly agree, Agree, Disagree, Strongly Disagree*." According to DSF results, US respondents need more ability to move across steps, for instance from "Strongly agree" to "Agree," than Spanish respondents with the same attitude toward science.

9.3.2.2 Cognitive Interviewing Results Themes and subthemes were identified from cognitive interview transcripts and then compared across groups in order to identify different interpretative patterns and processes. As a sample of some of the cognitive interviewing findings, it was noted that US and Spanish participants differed in their interpretation of the meaning of global expressions such as *Advances in broad science and technology*. Spanish respondents often conceptualized advances in limited terms, referring to everyday technologies such as mobile phones, the internet, and electrical appliances. On the other hand, US participants had broader, more expansive conceptualizations, such as alternative energy. Indeed, items with specific terms like "advances" and "inventions" were flagged with a large DIF, suggesting different interpretive patterns between US and Spanish respondents.

The mixed method design allowed researchers to interpret DIF by linking qualitative evidence from cognitive interviews to DIF statistics. The integration necessarily followed through all phases in the study, revealing not only the interpretative patterns found in the cognitive interviews, but also the extent to which the difference in the content and life experiences for US and Spanish respondents could undermine the validity of the cross-country comparisons. This case study shows how integrating cognitive interviews within a mixed research framework can provide a more complete picture of question performance. The next and final example illustrates how cognitive interviewing in mixed research can address data quality for psychometric scales of measurement.

9.3.3 Case 3: Cognitive Interviewing and Psychometric Scales

Validity theory in psychometrics has changed dramatically since the beginning of psychological testing in the early twentieth century (for recent work summarizing the Theory of Validity, see Kane 2013; Sireci and Padilla 2014). The latest edition available of the *Standards for Educational and Psychological Testing* (American Educational Research Association, American Psychological Association, National Council of Measurement in Education 1999) proposes a framework for validation studies based on five sources of validity evidence. One source of validity evidence incorporated in this edition is that of the evidence based on the response processes. According to the *Standards* (AERA et al. 1999), evidence based on response process refers to "evidence concerning the fit between the construct and the detailed nature of performance or response actually engaged in by examinees" (p. 12). The *Standards* (AERA et al. 1999) offer no clear indication regarding the methods for gathering validity evidence about the response processes. Cognitive interviewing studies can serve an important role answering current questions about empirical and theoretical analyses of the response processes.

Although the application of cognitive interviewing has been mainly related to survey questions, it can also be implemented for gathering validity evidence of psychological scales included in survey questionnaires (Padilla and Benítez 2014). Cognitive interviewing studies not only provide evidence of item performance, but also help to interpret traditional psychometrics like item-test correlation, reliability indexes and factor analysis. Findings about different interpretative patterns and processes can be associated with specific psychometrics for groups of respondents defined by cultural, linguistic or demographic variables.

On the other hand, the extent to which different interpretative patterns and processes uncovered by cognitive interviews affect survey data can be found by linking cognitive interviewing findings to psychometrics calculated on data from field tests, surveys, or both. Padilla, Benítez, and Castillo (2013) conducted a mixed research study in order to understand psychometrics of a psychological scale included in a national health survey and to measure a family function construct.

The study sought to determine whether validity evidence provided by cognitive interviews was useful to interpret psychometrics obtained from survey data. The APGAR scale was developed by Smilkstein (1978) to assess the construct of "family support" by evaluating the perception respondents have of the support they receive from family members. The scale consists of five polytomous items with three response categories. The APGAR scale is frequently included in health survey questionnaires due to its brevity and usefulness. Table 9.9 presents the original version of items of the APGAR scale.

The APGAR scale was developed to be applied individually by health professionals in a face-to-face interview as part of a health program or intervention. The Spanish version of the APGAR scale was included in the adult questionnaire of the Spanish Health Survey (Spanish Ministry of Health and Consumption 2006) administered to

TABLE 9.9 Original APGAR Items

	Almost always	Some of the time	Hardly ever
1. Are you satisfied with the help you receive from your family when you have a problem?	☐ 0	☐ 1	☐ 2
2. Are you satisfied with the time you and your family spend together?	☐ 0	☐ 1	☐ 2
3. Do you feel your family loves you?	☐ 0	☐ 1	☐ 2
4. Do you talk together about problems you have in home?	☐ 0	☐ 1	☐ 2
5. Important decisions are made by all of you together in home?	☐ 0	☐ 1	☐ 2

31,300 Spanish households. The mixed research agenda was posed as several related research questions. First, would there be different interpretative patterns and response processes to APGAR items between respondents living alone and those living with others? And if so, could it affect the APGAR scale for both groups?

To address these research questions, Padilla et al. (2013) planned a mixed research design that could be represented as QUAN+QUAL or quan+qual, following Creswell's (1995) classification. Both representations indicate that a study starts with a quantitative phase and finishes with the qualitative one. In addition, the classification shows that both phases were equally dominant.

The validation study was developed in two phases. First, the quantitative phase consisted of analysis of the psychometric properties of the scale based on survey data. Survey data were collected by interviewers at the Spanish Statistics Institute from June 2006 to June 2007. APGAR items were administered through face-to-face interviews, using the information collected from the household roster. Second, cognitive interviews were conducted in order to obtain (a) evidence of what APGAR items were capturing, and (b) possible different interpretation patterns between respondents living alone versus those living with others. Ten men and 11 women, aged 20–67 years, were recruited to take part in cognitive interviewing. Of these, 12 were single, 6 married, and 3 divorced. Nine people lived alone and 12 lived with other people.

9.3.3.1 Psychometric Results All analyses were conducted by dividing respondents according to the type of household (alone versus living with others). To sum up the main psychometrics, for all APGAR items, standard deviation values were significantly lower in the group of respondents living with others, indicating a lower variability in the responses among this group for each item. The corrected item-total correlation for the discrimination index (DI) of the items was within the typical values for these scales, with higher values in the group of respondents living alone. Cronbach's alpha coefficient was 0.83 for the group of single-person households and 0.68 for multi-person households. The dimensionality of responses to APGAR items in both groups was analyzed using exploratory factor analysis of principal axes. In both groups a single factor was obtained having an eigenvalue greater than one, and that accounted for 62.54% of the variance for respondents living alone, and 48.97% for those living with others. Table 9.10 shows the factor loadings of the items for one-factor solution of the groups.

TABLE 9.10 Factor loadings of the APGAR Items

	Factor loadings	
Items	Living alone	Living with others
1	0.80	0.67
2	0.81	0.73
3	0.74	0.64
4	0.61	0.35
5	0.75	0.64

Psychometrics showed that APGAR items can work differently in both groups. The percentage of explained variance and factor loadings were higher for the groups of respondents living alone than those living with others.

Quantitative analysis showed that a difference existed between these two types of households, but it could not explain why. Cognitive interviewing was incorporated into the research agenda in order to answer this question. These results are discussed next.

9.3.3.2 Cognitive Interviewing Results Evidence of three general themes was obtained: the concept of "family," "asking for help," and "making decisions." A comparative analysis was conducted to detect differences in the interpretative patterns and processes by household type (living alone versus living with others). First, differences were observed in the descriptions of members who were included in the family. Of the nine respondents living alone, most (7) responded to the items thinking about others that do not live with them, including their parents, children or siblings. One respondent even thought about his friends. Only one respondent stated "currently none, as I live alone." Of the respondents living with others, almost half (5) of the 12 responded to the items thinking only of the people with whom they lived; the other seven included family members living outside of the home.

Similar differences were also found with respect to the kind of help asked for and/or provided, and the people involved in the decision making processes. For example, the interpretation of the concept of "aid/help" for respondents living alone was more varied and heterogeneous than in the case of respondents living with others. In addition, respondents living alone more often understood the concept within the meaning of material need, compared to the emotional need for respondents living alone, indicating that the construct indicators could be different across the groups.

Differences in interpretative patterns and response processes could be associated with the findings of the psychometric analysis. Evidence from cognitive interviews was used to link different interpretations to the psychometric results, including (a) the greater variety in the interpretation of the types of assistance and the people who share or shared decision making, (b) the greater heterogeneity in the responses of the group of respondents from single-person households than in those who live with others and, as a result, (c) higher values of discrimination indices of the items, internal consistency of the scale (i.e., Cronbach's alpha coefficient of 0.83 for the single-person household group and 0.68 for the multi-person household group), and clearer unidimensionality (i.e., first factor accounted for 62.54% of the variance for respondents living alone, and 48.97% for those living with others). In addition, the difference in factor loading for item 4 ("*Do you talk together about problems you have at home?*") for both groups can be associated with differences that were observed in the descriptions of members who were included in the family.

Case 3 illustrates how cognitive interviewing findings can provide useful insights into response errors in survey data via psychometrics. Combining cognitive interviewing data and psychometrics in a mixed method study allowed Padilla et al. (2013) to understand psychometrics calculated on survey data for two groups of respondents. In addition, authors were able to obtain evidence of the response processes for groups defined by a variable (living alone versus living with others), which could have undermined validity of survey estimates if similar interpretative patterns and processes had been taken for granted.

9.4 CONCLUSION

In this chapter, the contents and main characteristics of the mixed research framework have been presented. In addition, the advantages of conducting cognitive interviews within a mixed research framework have been discussed. Mixed method studies that combine cognitive interviewing findings and results from other quantitative methods in order to address survey question evaluation and validation can improve the validity of survey estimates.

To address the lack of convergence between question evaluation methods is not only a matter of applying more than one method. Rather, a mixed research approach to survey question evaluation projects combines qualitative and quantitative methods at every stage of the research process, from formulating the objective, through the design, to concluding with the integration of the results. By combining cognitive interviewing studies with quantitative methodologies, understanding of question performance grows, leading to more informed decisions about whether a question should be altered, how the resulting variable should be utilized, and whether the survey estimate should be interpreted in a particular way.

10 Conclusion

KRISTEN MILLER and STEPHANIE WILLSON
National Center for Health Statistics

VALERIE CHEPP
Hamline University

JOSÉ-LUIS PADILLA
University of Granada, Spain

10.1 INTRODUCTION

The primary purpose of this book has been to put forward a methodological practice for conducting cognitive interviewing studies. By extension, the book conveys a broader picture of the advantages and benefits of performing cognitive interviewing studies within the context of survey research. Traditionally, the method has been used as a tool to identify question design "flaws." In this capacity, cognitive interviews provide survey managers an opportunity to pre-test questions—to see if respondents experience difficulty when answering their questions. If problems are identified, questions can presumably be "fixed" or "tweaked" before fielding. As a scientific methodology, however, cognitive interviewing studies provide a much more extensive and valuable contribution to the field.

Cognitive interviewing methodology provides an understanding of the way in which survey questions perform, specifically the phenomena (i.e., the construct) represented in resulting statistics. In that the method identifies the content or experiences considered in respondents' answers, it addresses construct validity. In that the method allows for analysis of interpretive patterns across groups, it addresses comparability, effective in determining accuracy of translations and equivalence across socio-cultural groups. When combined with quantitative methodology, cognitive interviewing studies provide a basis for understanding the types as well as the extent of various phenomena captured by a question. Knowledge from cognitive interviewing studies is of key importance for survey managers as it demonstrates that questions indeed capture the intended constructs. Moreover, when made public, findings from cognitive interviewing studies provide data users a more sophisticated understanding

Cognitive Interviewing Methodology, First Edition.
Edited by Kristen Miller, Stephanie Willson, Valerie Chepp, and José-Luis Padilla.

of survey data, allowing them to make better sense of their own research findings. Given the value, standardized and routine incorporation of cognitive interviewing methodology (and the publication of study findings) into the survey process would be a significant contribution toward the advancement of survey research, particularly in the area of construct validity and the reduction of specification error.

To serve this larger purpose, cognitive interviewing studies must adhere to the requirements of scientific methodology, that is, systematic and transparent analysis of empirical data. Without these attributes, findings become anecdotal and opinion-based and, ultimately, not useful to an empirically based field such as survey research. This book has outlined methodological practices and processes that support this type of study. The aim of this chapter is to summarize the significant practices and procedures of cognitive interviewing methodology articulated throughout this book. The chapter concludes by outlining future directions for the method as well as for survey research. Those directions include the study of cognitive interviewing methodology, expansion of mixed method research, and development of accepted standards for cognitive interviewing studies.

10.2 SUMMARY OF PRACTICES

The methodological practices put forward in this volume are specifically designed to understand the interpretive patterns used by respondents as they process questions and formulate answers. By understanding the various interpretations it is possible to better understand the construct captured by the survey question and represented by a survey statistic. The section below provides a summary of the practices described in this book.

10.2.1 Data Collection

Like all research, cognitive interviewing studies should begin with a methodological plan. This includes (but is not limited to) articulating the study's purpose and developing a sampling plan, interview guide, recruitment strategy, and plan for data collection. The methodological plan, its implementation, as well as any alterations, should be detailed in the study's final report.

A well-considered sampling plan is a necessary component of a cognitive interviewing study that will be able to provide a complete portrayal of a question's performance. Sample characteristics should be chosen to elicit the greatest diversity in regard to the various processes or patterns that respondents may use when answering a survey question.

Development of a sampling and recruitment plan. Cognitive interviewing employs purposeful sampling. This sampling method is used to ensure that the respondents have the characteristics necessary to provide relevant data, that is, data required to meet the study objectives. Because certain types of respondents are important to include in the study, a sampling and recruitment plan should be developed that includes the needed types of respondents. Recruitment may include advertisements in newspapers, flyers, community contacts, and word-of-mouth.

All conclusions to be made from a cognitive interviewing study depend upon on the composition and size of the sample. Decisions about whom to interview and how many people to interview should be guided by the following qualitative methodological principles:

(a) *Sample composition*: Respondents are identified based on their relationship to or experiences with the key characteristics of the study. The composition of the intended survey population and the homogeneity of the population's experiences with the key topics or variables of interest also play a role in determining the number of subgroups included in a cognitive interviewing study. If the respondent population is expected to be similar in their experiences and reaction to the key variables, there may not be a need for subgroups. If, on the other hand, there are known or suspected differences in the way in which particular populations may experience or interpret a construct, multiple subgroups would be desired. If subgroups are identified, the number of respondents within each subgroup is defined using the same principles described above; that is, include enough respondents to be able to identify, define and characterize themes in both question interpretation and response error.

(b) *Sample size*: A sample size goal may be decided at the onset of the study, but the final sample size is determined by the data being collected. Ongoing analysis of the data determines when saturation has been reached (i.e., when conceptual themes are fully developed and defined) and, therefore, informs when interviewing may cease, even if initial sample size has not been met. On the other hand, the sample size goal may be exceeded if the point of saturation has not been reached, or if additional issues or subgroups have been identified that need to be explored.

Development of an interview guide. The goal of the interview is to collect, in narrative format, the processes by which a respondent interprets and responds to a question as well as any difficulties experienced by a respondent in providing an answer. Respondents are placed in the role of storyteller. Specifically, they are asked to generate a narrative that depicts "why they answered the question in the way that they did." In telling the story, respondents relay what they believed the question was asking as well as describe the various factors that they considered in forming their answer. This narrative—in its complete form—details the particular context of the respondent's life and the various experiences they considered in order to arrive at their answer. The interview guide, then, should be structured to elicit this type of data. Typically, interview guides contain the questions to be evaluated along with follow-up questions or general ideas and topics to be explored.

In keeping with the goal of achieving narrative data, experience-based probes are used. The probes used to elicit a "storyteller mode" typically ask the respondent to speak from their own experience, describing and not evaluating their thoughts. Rather than asking respondents to be question design experts, experience-based probes assume only that the respondent is an "expert" on their own personal experiences, that is, they can only serve as informants to their own experience. As such, these types of probes are follow-up questions that ask respondents to *describe and explain* this experience. This yields data that convey authenticity and are convincing in terms of why respondents answered the survey questions the way they did.

In some instances, respondents may be asked experience-based probes, but continue to give their opinions about the question as well as speculate about the problems that other respondents may have. It is incumbent upon the interviewer to recognize this type of information and redirect the respondent through another line of questioning to achieve experience-based information.

Role of the interviewer. When the target of investigation is conceived as a study of meaning and interpretation as it pertains to life experience, the role of the interviewer shifts to that of a qualitative researcher. The interview is understood as a complex interaction, in which the interviewer plays an integral and active role. The interviewer must, in the course of the interview, assess the information that he or she is collecting and examine the emerging information to identify any gaps, contradictions or incongruences in the respondent's narrative. The interviewer, then, uses his or her analytic skills to form additional probe questions so that a complete and coherent story is garnered.

10.2.2 Analysis

Systematic and transparent analyses of empirical data are key to producing valid and reputable cognitive interviewing findings. A systematic analysis ensures that no one particular case is over-emphasized and that findings do not appear anecdotal. While one particular case may be highly relevant and require particular attention, discounting other cases produces incomplete or misleading findings. The notion of transparency is critical in that it allows readers to understand as well as to examine the ways in which analyses were conducted and how conclusions were reached. Transparency instills the trustworthiness of a study and the reputability and believability of its findings. These tenets carry through data collection and analysis to the final report, which must document the analytic process and present evidence to support findings.

An essential component of a methodological plan includes a plan for analysis. Cognitive interviewing studies are based on empirical data collected from respondent interviews. As a qualitative method it generates textual data derived from in-depth interviews. Raw data of a cognitive interviewing study consists of either a video or audio recording or a written transcript of the interview. As is the case for all analyses of qualitative data, the general process involves data synthesis and reduction—beginning with a large amount of textual data and resulting in conclusions that are meaningful to the ultimate purpose of the study. Analysts must examine data within interviews, across interviews (by question) and across subgroups with the goal of identifying thematic patterns in question interpretations and response errors. The process can be conceptualized within five incremental steps.

1. Conducting interviews, collecting the ways in which a respondent interpreted and formulated answers to the survey questions,

2. Synthesizing interview text into summaries, detailing how respondents formulated their answers, including events or experiences considered as well as any difficulties answering the question,
3. Comparing summaries across respondents to identify common themes and to develop a thematic schema that details phenomena captured,
4. Comparing those themes across subgroups to identify ways in which different groups may process questions differently depending on their differing experiences and socio-cultural backgrounds,
5. Making conclusions based on the thematic schema that depict how each question performs as well as providing explanation for the performance.

In developing thematic schema, the analyst determines whether an interpretive pattern has emerged across respondents. These decisions are significant as they directly bear upon the ultimate findings of the cognitive interviewing study. For this reason, it is important to consider the ways in which such decision making can be made transparent so that results can be replicated or, at least, understood. An audit trail, consisting of the analytic products generated from each level of analysis, reveals the types of decisions made by the analyst in order to conduct the study.

To fulfill the highest degree of transparency, several steps should be taken. The respondent–interviewer interaction generated from the interview process is considered raw data. Thus, interviews should be documented by audio or video recordings. Transcriptions may follow, but at the very least there should be a saved recording of the interview. Researchers should also create an audit trail which illustrates how data were synthesized. This can be performed with such applications as Q-Notes designed specifically for the five-tier analysis of cognitive interviews or another qualitative software application such as NVivo. At the very least, an Excel spreadsheet or Word file can be used. Finally, the study should be documented with a final report. Complete reporting improves the rigor and credibility of the study, maximizes the transparency and repeatability of the analyses, and is essential for evaluating study quality.

10.2.3 Documenting Study Findings

Conveying research results is a necessary part of the scientific process because it not only documents the findings, but also how they were attained. This includes a detailed discussion of how data are collected and analyzed and how conclusions are drawn. By documenting research findings and making them available to a wider public, it is possible to assess the quality of the empirical data produced as well as the veracity of the findings.

The aim of cognitive interviewing reports is to present findings pertaining to the performance of survey questions. By presenting these findings, as well as providing information on methodological issues raised or discovered during a testing project, cognitive interviewing reports also help readers obtain specific information about how a new question evaluation project can be designed in order to build on the results of previous projects. Depending on the goal(s) of the cognitive interviewing study, reports can focus on different aspects of question performance, including difficulties respondents experienced when answering a question, a question's construct validity, or issues of question comparability (e.g., comparability across socio-cultural contexts or translations). Cognitive testing of every language version that will be fielded is best practice. It is then possible to answer whether a question is being interpreted as intended in every language in which data will be collected.

There is no single way to structure a cognitive interviewing report. However, all cognitive interviewing reports should consist of a basic series of five clearly marked sections: an introduction, a summary of the findings, a description of the methods, a detailed question-by-question review, and an appendix. Each section should be concise, yet thorough. This organizational format results in a document that is easy to read and accessible across different audiences.

I. *Introduction*. This section orients the reader to the study. It includes background information about the survey or topic and offers a statement of the study purpose. In addition, it lays out the rationale for using cognitive interviewing as the best methodological choice, given the research question(s).

II. *Summary of Findings*. This section highlights the study's main conclusions and synthesizes the findings across all the questions.

III. *Methods*. This section of a report is a description of the procedures associated with the method. It includes, for example, a description of sampling frame and recruiting method, with a discussion of both the target sample (number and characteristics) in comparison to the achieved sample (number and characteristics). The methods section also has a description of the interview protocol (i.e., the topics and issues to be covered by the interviewer) and the interview guide itself, often included as an attachment. Details on the interviewer(s), interview location, interview length, etc. are also discussed, along with the method of note taking, session recording and/or use of transcripts. Finally, a discussion of any changes to methods during testing (and why) is included in this section, as well as a discussion of study limitations (e.g., sample issues) and reflexivity (e.g., any interactional effects between the interviewer and respondent that may have compromised data quality).

IV. *Question-by-Question Review*. This section should discuss the results of the study in a detailed manner. This includes a general overview of responses and an explication of the key findings and themes for each question, complete with illustrative quotes and examples. Specific patterns and trends should also be identified. The specific patterns found should be linked to the summary of findings as a way for the reader to see how the general findings are grounded in empirical data. Finally, if appropriate, a discussion and explanation of counterexamples should be included and related to the main patterns and larger themes for each question.

V. *Appendix*. A final report should include the interview protocol and any supplemental methodological documentation, typically as an addendum.

Just as all cognitive interviewing reports should follow a basic structure, reports must establish credibility. Credibility, or trustworthiness, refers to the validity of a qualitative study. The credibility of research results relies primarily on the trust the audience has in the researcher. In cognitive interviewing reports, like other qualitative methods, credibility is demonstrated by a researcher's transparency and reflexivity. The writer demonstrates transparency by providing a clear research audit trail that details methods and procedures and reflexivity by being aware of the influences they have on research processes.

10.3 NEW DIRECTIONS

Since the 1980s, cognitive interviewing for survey question evaluation has evolved significantly. Throughout the years various aspects of the method have been discussed and debated, for example, the advantages of retrospective verses concurrent probing, the degree of interview structure (i.e., the amount of pre-scripted probes), and the appropriate number of respondents. This book aims to spur new topics of discussion with the goal of methodological advancement. As a qualitative methodology, cognitive interviewing is uniquely able to examine construct validity and specification error. However, in order to fully benefit from cognitive interviewing studies, the method itself requires examination with methodologists engaging in careful self-reflection.

10.3.1 Topics for Examination

The way in which the phenomena of any study is conceptualized guides the ways in which empirical research is conducted, as well as why it is conducted in the first place. It also provides insight into why some methods are more appropriate for specific types of research questions than others. This book is set within an interpretivist framework in which the construction of meaning is seen as elemental to the question-response process and is the phenomena of examination in a cognitive interviewing study. The method explicated in this book, then, is oriented toward the collection and analysis of interpretive patterns and processes that constitute the question-response processes. As indicated in earlier chapters, this perspective does not run counter to the psychological focus of cognition, but rather emphasizes interpretive value and the fluidity of meaning within the context of a questionnaire, as well as within the socio-cultural context of respondents' lives.

In moving forward with this discussion, the framework put forth in this book presents additional questions to be addressed. For example: In what ways does an interpretivist framework differ from a cognitive psychological framework? In what ways do they overlap? Perhaps more importantly, are there better ways to conceptualize the question-response process so that it is understood more completely? Or, is there another perspective that provides a different type of understanding? If so, what is the best method for examining those processes?

Another important issue requiring consideration is study of the method itself. In the past, the value of cognitive interviewing studies has been assessed by a variety of factors: the number of "problems" detected, the "real-ness" of the problems detected, and the ability to correct discovered problems and to produce problem-less questions

(Conrad and Blair 2009). If, however, the goal of a cognitive interviewing study is to discover patterns of interpretation—while understanding that interpretations are not static or monolithic but dependent on respondent experience and social location—these types of assessments are not appropriate. Since the unit of analysis is not "flaws" in questions but rather patterns of interpretation, the more appropriate assessments should focus on the completeness or "saturation" of the identified patterns. Are all of the possible ways that respondents would interpret a question discovered? What types of patterns are not identified? And, why not? Were respondents with particular characteristics not selected that should have been included in the sample? In addition, are the relationships between patterns adequately understood to provide a complete picture of question performance? Studies that are designed to answer these types of overarching questions will contribute to the development of cognitive interviewing methodology.

Other investigations into the method should pertain to data quality since the interview data themselves directly bear upon study findings. Qualitative literature is full of discussion pertaining to data quality that could provide useful guidance for examining cognitive interviewing data quality. For example, the insider versus outsider debate has been an important discussion within the qualitative methodology community. To what extent does the race or gender of the interviewer and the respondent impact data quality? What other characteristics of sameness or difference between the interviewer and the respondent impact the cognitive interview? Other issues to examine could be the effects of the interview setting on data quality, for example, in an office building or in respondents' homes. Finally, the data collection chapter of this book had a rather in-depth discussion of experience-based probes as opposed to evaluative-probes. However, more work is needed to refine and advance these ideas. For example, what other types of interviewer questioning exist and how does this impact the type and quality of data?

10.3.2 Mixed Method Research

It is increasingly recognized that a more complete understanding of question performance is achieved with a mixed research approach to question evaluation. Cognitive interviewing methodology has limitations as does every other methodology. Designing a study that capitalizes on the strengths of multiple methods represents one of the most fruitful paths in question design and survey data quality. Mixed method research is an excellent tool for achieving a more complete understanding of survey question performance, construct validity of survey measures, and the empirical meaning of estimates.

The mixed method chapter of this book presents three different examples of studies that combine cognitive interviewing with a quantitative methodology. All three examples illustrate how cognitive interviewing, as a qualitative methodology, is uniquely able to depict the interpretive meanings behind survey statistics. In addition, the examples illustrate how the quantitative methodologies supplement cognitive interviewing findings with a statistical understanding. These examples, however, are only illustrations of the potential for mixed method design that utilizes cognitive interviewing methodology. All quantitative question evaluation methodologies would benefit from the interpretive value of a cognitive interviewing study. For example, a split sample experiment in which results show that one question attains higher estimates than another question worded differently cannot in itself determine which estimate is "correct." Findings from a cognitive interviewing study, however, could provide insight such that this determination could be made. Research toward the advancement of mixed method designs will contribute to a better understanding of response error and, ultimately, how it impacts the total survey error.

10.3.3 Accepted Standards of Cognitive Interviewing Studies

Another avenue for advancing the methodology of cognitive interviewing is to develop accepted standards. The field of survey methodology has many forms of best practices. For example, standards and guidelines for collecting and producing survey data have been established for the US federal statistical system (Office of Management and Budget 2006). Those guidelines include such practices regarding sample and calculating non-response. Best practices raise awareness of the types of issues that can undermine the validity of the resulting data, and provide strategies for overcoming potential hazards.

Perhaps more significantly to the field of survey methodology, best practices establish a baseline for the field in understanding the ultimate product, that is, the survey data. Thus, when a survey can point to an agreed upon set of standards and indicate that these were the followed procedures, researchers can understand the specific value and limitations of the data. In order for cognitive interviewing studies to be utilized in a similar fashion it is necessary for this methodology to move toward similar practices. A set of standards or guidelines should not be considered static or unamendable. To the contrary, those principles should develop and change as methodological knowledge advances.

It is the intention of this book to move the field in this direction. It has shown how cognitive interviewing studies can be utilized as a methodology with unique benefits to the field of survey research. The aim of this work is to stimulate meaningful debate and discussion within the community of question evaluation methodologists. In addition, the goal of this book is to illustrate the full benefits and usefulness of the method to survey managers, methodologists and data users. As a scientific methodology that involves a systematic and transparent analysis of empirical data, the method becomes indispensable to all participants of survey research.

KEY CONCEPTS

Audit Trail: Documentation of the decisions made by the researcher at each stage of the research process. An audit trail allows readers to evaluate the veracity of conclusions drawn in the study.

Cognitive Aspects of Survey Methodology (CASM): The interdisciplinary science involving the intersection of cognitive psychology and survey methods. CASM research focuses on the mental processing of respondents and how those processes inform the survey response process.

Cognitive Sociology: A subfield of sociology that theoretically and empirically demonstrates how cognitive processes are shaped by cultural phenomena.

Constant Comparative Method: Often associated with the seminal work of Glaser and Strauss (1967), this is a method for analyzing data in order to develop a grounded theory. It is an inductive and iterative process in which the analyst continuously moves back and forth from raw data text, themes, and emerging conceptual claims. In cognitive interviewing, the analyst compares data across survey questions, both within a single respondent interview and among all interviews, in order to arrive at a complete understanding of the question-response process and the construct measured by the question.

Construct Validity: The extent to which a survey question captures what it was intended to measure.

Credibility: Also called trustworthiness, this refers to the validity of qualitative research, including cognitive interviews. A study is credible to the extent that an audience trusts the accuracy of the conclusions.

Data Reduction: The process by which an analyst groups large amounts of data into a smaller number of meaningful categories that summarize and explain the question-response process and the construct measured by the question.

Decentering: An approach to multilingual translation in which text is translated back and forth between two (and potentially more) languages until all versions are found to be equivalent.

Four-Stage Cognitive Model: A theoretical model detailing each stage of the question-response process, typically including: (1) comprehension, (2) recall, (3) judgment, and (4) response. The model is often used to guide cognitive interviewing techniques in order to uncover problems in the question-response process.

Cognitive Interviewing Methodology, First Edition.
Edited by Kristen Miller, Stephanie Willson, Valerie Chepp, and José-Luis Padilla.

Grounded Theory: An approach used to generate theories that explain how a particular phenomenon operates. The approach is based on inductive reasoning, building theoretical claims directly from empirical observation. In cognitive interviewing, the approach is used to develop explanations of survey question performance.

Interpretivism: An intellectual tradition emphasizing the role of meaning in individuals' engagement with the social world. Cognitive interviewing methodology uses this approach to focus on respondents' interpretations of survey questions and on how their lived experiences inform their answers, with special attention paid to the socio-cultural context of the question-response process.

Narrative: A way of organizing and making sense of experiences by putting those experiences into a structured sequence of events with a beginning, middle, and end. As a cognitive interviewing technique, this is a form of storytelling in which individuals are asked to convey their understanding of themselves, their life histories and personal circumstances, as well as their interpretations of survey questions.

Probe: A cognitive interviewing technique in which an interviewer asks follow-up questions of the respondent in order to understand the question-response process and the construct being measured by the question.

Purposive Sample: A non-random, theoretically driven sample that is deliberately selected to achieve a particular goal.

Question-Response Process The cognitive and social processes in which respondents engage when answering survey questions. Cognitive interviewers seek to replicate this experience through the cognitive interview in order to understand the processes by which respondents answer survey questions.

Reflexivity: A core component of qualitative research, the process by which researchers reflect on and document the ways in which their presence and line of questioning shape data.

Social Location: The way in which individuals and groups are differently located within a social structure, based upon social markers such as race, ethnicity, gender, social class, sexuality, and disability status, among others. Social location shapes individuals' and groups' experiences, worldviews and interpretations of survey questions.

Theoretical Relevance: The criterion governing the composition of a sample for a qualitative study. Respondents are chosen based on characteristics that will generate complete understanding of the study phenomena.

Theoretical Saturation: The criterion governing the size of a sample for a qualitative study. A cognitive interviewing study is complete when patterns of question interpretation are sufficiently understood and explicated.

Thematic Schema: Patterns of question interpretation that are analytically identified to explain the question-response process and the construct being measured by the survey question.

Thick Description: A common qualitative technique that cognitive interviewers can use to elicit from respondents a rich and detailed explanation of how and why they answered a survey question in a certain manner.

Think-Aloud: An approach used by cognitive interviewers to collect information regarding the cognitive aspects of the question-response process. With this approach, respondents verbalize their thought processes as they answer a survey question.

Transparency: This core feature of the scientific process presents research findings in a way that clearly details all methods and procedures followed throughout the research study. Researchers demonstrate transparency by providing a clear research audit trail documenting how the research was conducted from start to finish.

TRAPD Model: A team translation model where two or more individuals translate the survey instrument into the desired target language. The translators and at least one reviewer then meet to review the original translation(s) and make comments on issues they find or changes they recommend. In this model, decisions made at every step are documented to inform designers and analysts about how the final translation was reached.

QUESTION EVALUATION RESOURCES

Below is a sampling of resources pertaining to cognitive interviewing methodology, question evaluation, and measurement error.

ONLINE RESOURCES

Q-Bank

Website: http://wwwn.cdc.gov/QBank

Q-Bank is an online resource that houses cognitive interviewing and other question evaluation reports. The database is searchable by question and links each question to test findings. Q-Bank allows data analysts to better understand survey data regarding the interpretive value of the question. Q-Bank fosters the sharing of information among survey researchers, providing the ability to compare evaluation studies and to examine where certain methods can be improved. Q-Bank is a collaborative effort of many federal agencies and contractors, including the National Center for Health Statistics, the US Bureau of the Census, National Cancer Institute of the National Institutes of Health, Bureau of Labor Statistics, the National Science Foundation, and other statistical agencies.

Q-Notes

Website: http://wwwn.cdc.gov/qnotes

Q-Notes is a qualitative research tool developed by the National Center for Health Statistics to assist in the management and analysis of cognitive interviewing studies. Q-Notes is available online and provides interviewers and analysts real-time access to interview data. This online application allows interviews to be conducted in multiple geographical regions so that comparability can be examined for multi-national and multi-lingual surveys. In addition, the tool allows project managers to easily monitor the status of their projects. The analytic features of Q-Notes allow for a more rapid, yet thorough and systematic analysis of the data utilizing the analytic procedures outlined

Cognitive Interviewing Methodology, First Edition.
Edited by Kristen Miller, Stephanie Willson, Valerie Chepp, and José-Luis Padilla.

in this book. Q-Notes provides simple ways to conduct simultaneous analytical steps of data reduction and knowledge production. Q-Notes is currently used by numerous statistical agencies around the world both for their own projects and to collaborate with other agencies internationally. Q-Notes is managed by the Question Design Research Program at NCHS and is continually updated to provide users with the most benefits.

Cross-Cultural Survey Guidelines

Website: http://ccsg.isr.umich.edu/index.cfm

"These guidelines were developed as part of the Comparative Survey Design and Implementation (CSDI) Guidelines Initiative. The aim of the Initiative was to promote internationally recognized guidelines that highlight best practice for the conduct of comparative survey research across cultures and countries. The intended audience is researchers and survey practitioners planning or engaged in cross-cultural or cross-national research." The website is hosted by the University of Michigan Institute for Social Research.

CHECKLISTS AND STANDARDS

Cognitive Interviewing Reporting Format Checklist:

The Cognitive Interviewing Reporting Format (CIRF) is a guide to the reporting of cognitive interviewing results. The CIRF was designed as a checklist of 10 major elements to be described within a cognitive interviewing report (Research Objectives, Research Design, Ethics, Participant Selection, Data Collection, Data Analysis, Findings, Conclusions, Strengths and Limitations, and Report Format). The goals of the CIRF are to lead practitioners of cognitive testing to provide comprehensive, transparent documentation of the complete set of steps involved in conducting their research, and to create the basis for a harmonized set of reporting standards. A 2013 special issue of the journal *Methodology* (Willis and Boieje, Volume 9(3)) was devoted to the introduction, description, and initial application of the CIRF.

ACADEMIC JOURNALS

"***Field Methods*** is an indispensable tool for scholars, students, and professionals who do fieldwork. It offers important refereed articles, descriptions of methodological advances, advice on the use of specific field techniques, help with both qualitative and quantitative methods—all the tools necessary for those who conduct fieldwork." A 2011 special issue of Field Methods (Willis and Miller, 23(4)) was devoted to cross-cultural cognitive interviewing (http://fmx.sagepub.com/).

"***International Journal of Public Opinion Research*** is a source of informed analysis and comment for both professionals and academics. Edited by a board drawn from over a dozen countries and several disciplines, and operated on a professional referee system, the journal is the first truly comparative, multidisciplinary forum serving the international community. In addition to original articles, the journal includes review articles, surveys of recent developments in the field, major book reviews and notes section, journal abstracts, information about forthcoming conferences, and news about its sponsoring body, the World Association for Public Opinion Research (WAPOR)" (http://ijpor.oxfordjournals.org/).

"***Journal of Official Statistics*** **(JOS)** publishes articles on statistical methodology and theory, with an emphasis on applications, and although the journal is focused on the production of federal-level statistics, its scope includes methodological developments that apply widely to the survey field. JOS encompasses the full range of issues related to survey error, and to survey quality generally. JOS is published by Statistics Sweden" (http://www.researchgate.net/journal/0282-423X_Journal_of_official_statistics).

"***Journal of Survey Statistics and Methodology*** is sponsored by AAPOR and the American Statistical Association. Its objective is to publish cutting-edge scholarly articles on statistical and methodological issues for sample surveys, censuses, administrative record systems, and other related data. Topics of interest include survey sample design, statistical inference, nonresponse, measurement error, the effects of modes of data collection, paradata and responsive survey design, combining data from multiple sources, record linkage, disclosure limitation, and other issues in survey statistics and methodology. The journal will publish both theoretical and applied papers, provided the theory is motivated by an important applied problem and the applied papers report on research that contributes generalizable knowledge to the field. Review papers are also welcomed. Papers on a broad range of surveys are encouraged, including (but not limited to) surveys concerning business, economics, marketing research, social science, environment, epidemiology, biostatistics, and official statistics" (http://jssam.oxfordjournals.org/).

"***Public Opinion Quarterly*** is among the most frequently cited journals of its kind. Such interdisciplinary leadership benefits academicians and all social science researchers by providing a trusted source for a wide range of high-quality research. POQ selectively publishes important theoretical contributions to opinion and communication research, analyses of current public opinion, and investigations of methodological issues involved in survey validity—including questionnaire construction, interviewing and interviewers, sampling strategy, and mode of administration. The theoretical and methodological advances detailed in pages of POQ ensure its importance as a research resource" (http://poq.oxfordjournals.org/).

"***Quality and Quantity*** constitutes a point of reference for European and non-European scholars to discuss instruments of methodology for more rigorous scientific results in the social sciences. The journal publishes papers on models of classification, methods for constructing typologies, models of simulation, neural networks and Fuzzy sets for social research, mathematical models applied to social mobility, mathematical models of voting behavior, qualitative methodology and feminist methodology, discussions on the general logic of empirical research, analysis of the validity and verification of social laws, and similar topics. Essentially, *Quality and*

Quantity is an interdisciplinary journal which systematically correlates disciplines such as mathematics and statistics with the social sciences, particularly sociology, economics, and social psychology. The ultimate aim of the journal is to widen the discussion of the most interesting contributions to methodology to scholars of different nations, the objective being the scientific development of social research" (http://www.springer.com/social+sciences/journal/11135).

PROFESSIONAL ORGANIZATIONS/MEETINGS

American Association for Public Opinion Research (AAPOR) is a professional organization dedicated to advancing the science and practice of survey and opinion research to give people a voice in the decisions that affect their daily lives. The organization hosts an annual conference where issues and new research on question evaluation are discussed, and workshops on cognitive interviewing are offered.

European Survey Research Association (ESRA) was founded in 2008 to facilitate communication and interactions between European survey researchers and with colleagues in non-European countries. ESRA organize a bi-annual conference and publish a peer-review journal call *Survey Research Methods*.

The World Association for Public Opinion Research (WAPOR) is a professional association of survey researchers from universities and private institutions. It is open to individual people who have an interest in the conduct, use, or teaching of scientific opinion and attitude surveys, social science research, media or communications research, market research, or related activities. The organization has approximately 500 members from almost 60 different countries. It was founded in 1947. WAPOR aims at organizing and sponsoring meetings and publications, encouraging high professional standards, promoting improved research techniques, informing journalists about the appropriate forms of publishing poll results, observing the democratic process, and use of polls in elections, promoting personnel training, and coordinating international polls.

REFERENCES

Ackermann AC, Blair J. (2006). Efficient respondent selection for cognitive interviewing. *Paper presented at the American Association of Public Opinion Research Annual Meeting*. Montreal, Canada.

Acquadro C, Jambon B, Ellis D, Marquis P. (1996). Language and translation issues. In: B Spilker (ed) *Quality of Life and Pharmacoeconomics in Clinical Trials*, 2nd edition. Philadelphia: Lippincott-Raven. pp. 575–585.

AERA, APA, NCME. (1999). *Standards for Educational and Psychological Testing*. Washington, DC: American Educational Research Association.

Alexander JC. (1987). *Twenty Lectures Sociological Theory Since World War II*. New York: Columbia University Press.

Alexander JC, Smith P. (2004). The strong program in cultural sociology. In: JC Alexander (ed) *The Meanings of Social Life*. New York: Oxford University Press. pp. 11–26.

Beatty P. (2004). Paradigms of cognitive interviewing practice and their implications for developing standards of best practices. *Proceedings of the 4th Conference on Question Evaluation Standards*. Zentrum fur Umfragen, Methoden Und Analysen (ZUMA).

Beatty P, Willis G. (2007). Research synthesis: the practice of cognitive interviewing. *Public Opinion Quarterly*, 71(2), 287–311.

Benítez I, Padilla JL. (2014). Analysis of non-equivalent assessments across different linguistic groups using a mixed methods approach: understanding the causes of differential item functioning by cognitive interviewing. *Journal of Mixed Methods Research*, 8(1), 52–68.

Berger P, Luckmann T. (1966). *The Social Construction of Reality*. Garden City, NJ: Doubleday.

Biemer PP. (2011). Total survey error: design, implementation, and evaluation. *Public Opinion Quarterly*, 74(5), 817–848. http://poq.oxfordjournals.org/cgi/doi/10.1093/poq/nfq058 (accessed March 15, 2013).

Blair J, Brick PD. (2010). Methods for the analysis of cognitive interviews. *Proceedings of the Section on Survey Research Methods*. Washington, DC: American Statistical Association.

Blair J, Conrad F, Ackermann A, Claxton G. (2006). The effects of sample size on cognitive interview findings. *Paper presented at the American Association of Public Opinion Research annual meeting*. Montreal, Canada.

Blanden AR, Rohr RE. (2009). Cognitive interview techniques reveal specific behaviors and issues that could affect patient satisfaction relative to hospitalists. *Journal of Hospital Medicine*, 4(9), E1–E6.

Cognitive Interviewing Methodology, First Edition.
Edited by Kristen Miller, Stephanie Willson, Valerie Chepp, and José-Luis Padilla.

Blumer H. (1969). *Symbolic Interactionism: Perspective and Method.* New Jersey, NJ: Prentice-Hall, Inc.

Boeije H, Willis G. (2013). The cognitive interviewing reporting framework (CIRF): towards the harmonization of cognitive testing reports. Methodology. *European Journal of Research Methods for the Behavioral and Social Sciences,* 9(3), 87–95.

Bolton R. (1991). An exploratory investigation of questionnaire pretesting with verbal protocol analysis. In: RH Holman, MR Solomon (eds) *Advances in Consumer Research,* Volume 18. Provo, UT: Association for Consumer Research. pp. 558–565.

Braun M, Harkness JA. (2005). Text and context: challenges to comparability in survey questions. In: JHP Hoffmeyer-Zlotnik, JA Harkness (eds) *Methodological Aspects in Cross-National Research (ZUMA-Nachrichten Spezial 11).* Mannheim, Germany: ZUMA. pp. 95–107. http://www.gesis.org/fileadmin/upload/forschung/publikationen/zeitschriften/zuma_nachrichten_spezial/znspezial11.pdf (accessed July 8, 2013).

Brekhus W. (2003). *Peacocks, Chameleons, Centaurs: Gay Suburbia and the Grammar of Social Identity.* Chicago, IL: University of Chicago Press.

Brekhus W. (2007). The Rutgers School: a Zerubavelian culturalist cognitive sociology. *European Journal of Social Theory,* 10(3), 448–464.

Bruner J. (1986). *Actual Minds, Possible Worlds.* Cambridge, MA: Harvard University Press.

Cast AD. (2003). Identities and behavior. In: PJ Burke, TJ Owens, RT Sherpe, PA Thoits (eds) *Advances in Identity Theory and Research.* New York: Kluwer Academic/Plenum Publishers. pp. 41–53.

Catania, JA, Binson D, Canchola J, Pollack LM, Hauck W, Coates TJ. (1996). Effects of Interviewer Gender, Interviewer Choice, and Item Wording on Responses to Questions Concerning Sexual Behavior. *Public Opinion Quarterly,* 60(3), 345–375.

Charmaz K. (2006). *Constructing Grounded Theory: A Practical Guide Through Qualitative Analysis.* Thousand Oaks, CA: Sage Publications.

Chi M. (1997). Quantifying qualitative analyses of verbal data: a practical guide. *The Journal of Learning Sciences,* 6(3), 271–315.

Collins P, Chepp V. (2013). Intersectionality. In: G Waylen, K Celis, J Kantola, L Weldon (eds) *The Oxford Handbook of Gender and Politics.* New York: Oxford University Press. pp. 57–87.

Comparative Survey Design and Implementation Workgroup (2011). Cross-Cultural Survey Guidelines (CCSG). http://ccsg.isr.umich.edu/index.cfm (accessed on April 17, 2014).

Conrad F, Blair J. (1996). From impressions to data: increasing the objectivity of cognitive interviews. *Proceedings of the Section on Survey Research Methods.* Washington, DC: American Statistical Association.

Conrad F, Blair J. (2004). Data quality in cognitive interviews: the case for verbal reports. In: S Presser, J Rothgeb, M Couper, J Lessler, E Martin, J Martin, E Singer (eds) *Methods for Testing and Evaluating Survey Questionnaires.* Hoboken, NJ: John Wiley & Sons.

Conrad FG, Blair J. (2009). Sources of error in cognitive interviews. *Public Opinion Quarterly,* 73, 32–55.

Cooley CH. (1902). *Human Nature and the Social Order.* New York: Charles Scribner's Sons.

Cortés DE, Gerena M, Canino G, Aguilar-Gaxiola S, Febo V, Magaña C, Soto J, Eisen SV. (2007). Translation and cultural adaptation of a mental health outcome measure: the BASIS-R(c). *Culture, Medicine and Psychiatry,* 31(1), 25–49.

Creswell JW. (1995). *Research Design: Qualitative and Quantitative Approaches.* Thousand Oaks, CA: Sage Publications.

Creswell JW. (1998). *Qualitative Inquiry and Research Design; Choosing Among Five Traditions.* Thousand Oaks, CA: Sage Publications.

DeGloma T, Friedman AM. (2005). Thinking with Socio-Mental Filters: Exploring the Social Structuring of Attention and Significance. *Paper presented at the annual meeting of the American Sociological Association.* Philadelphia, August 2005.

DeMaio TJ, Rothgeb J. (1996). Cognitive interviewing techniques in the lab and in the field. In: N. Schwartz, S. Sudman (eds) *Answering Questions: Methodology for Determining Cognitive and Communicative Processes in Survey Research.* San Francisco: Jossey-Bass. pp. 177–195.

DeMaio T, Ciochetto S, Davis W. (1993). Research on the Continuing Survey of Food Intakes by Individuals. *Proceedings of the Section on Survey Research Methods, 1021-6.* Washington, DC: American Statistical Association.

Denzin N. (1994). The art and politics of interpretation In: K Norman, K Denzin, YS Lincoln (eds) *Handbook of Qualitative Research.* Thousand Oaks, CA: Sage Publications.

Dey I. (1999). *Grounding Grounded Theory.* San Diego, CA: Academic Press.

DiMaggio P. (1997). Culture and cognition. *Annual Review of Sociology*, 23, 263–287.

Dorer B. (2013). Enhancing the Translatability of the Source Questionnaire in the European Social Survey (ESS) – Does Advance Translation Help? *Paper presented at the Annual Conference of the American Association for Public Opinion Research (AAPOR)*, May 16–19, 2013, Boston, MA.

Durkheim E. ([1912] 1995). *The Elementary Forms of the Religious Life.* New York: The Free Press.

Ellis C, Bochner AP. (2000). Autoethnography, personal narrative, reflexivity. In: NK. Denzin, YS. Lincoln (eds) *Handbook of Qualitative Research*, 2nd edition. Thousand Oaks, CA: Sage Publications. pp. 733–768.

Emerson R. (2001). *Contemporary Field Research: Perspectives and Formulations*, 2nd edition. Prospect Heights, IL: Waveland Press, Inc.

Ericsson K, Simon H. (1980). Verbal reports as data. *Psychological Review*, 87, 215–257.

Ericsson K, Simon H. (1993). *Protocol Analysis: Verbal Reports as Data.* Cambridge, MA: MIT Press.

European Social Survey. (2012). ESS Round 6 Translation Guidelines. Mannheim, European Social Survey GESIS. http://www.europeansocialsurvey.org/index.php?option=com_docman&task=doc_download&gid=943&Itemid=80 (accessed July 8, 2013).

Fitzgerald R, Widdop S, Gray M, Collins D. (2009). Testing for equivalence using cross-national cognitive interviewing. Centre for Comparative Social Surveys Working Paper Series, Paper no. 01. http://www.city.ac.uk/__data/assets/pdf_file/0014/125132/CCSS-Working-Paper-No-01.pdf (accessed July 8, 2013).

Forsyth B, Lessler J. (1991). Cognitive laboratory methods: a taxonomy. In: P Biemer, R Groves, L Lyberg, N Mathiowetz, S Sudman (eds) *Measurement Error in Surveys.* New York: John Wiley & Sons.

Franzosi R. (1998). Narrative analysis—or why (and how) sociologists should be interested in narrative. *Annual Review of Sociology*, 24, 517–554.

Geertz C. (1973). *Thick Description: Toward an Interpretative Theory of Culture. In the Interpretation of Cultures.* New York: Basic Books.

Gerber ER. (1999). The view from anthropology: ethnography and the cognitive interview. In: MG Sirken, DJ Herrmann, S Schechter, N Schwarz, JM Tanur, R Tourangeau (eds) *Cognition and Survey Research*. New York: John Wiley & Sons, Inc.

Gerber ER, Wellens TR. (1997). Perspectives on pretesting: 'cognition' in the cognitive interview? *Bulletin de Méthodologie Sociologique*, 55, 18–39.

Giddens A. (2001). Introduction. In: M. Weber (ed) *The Protestant Ethic and the Spirit of Capitalism*. New York: Routledge.

Gilgun J. (2010). Reflexivity and qualitative research. *Current Issues in Qualitative Research*, 1(2), 1–8. http://www.scribd.com/doc/35787948/Reflexivity-and-Qualitative-Research (accessed February 28, 2013).

Glaser B, Strauss A. (1967). *The Discovery of Grounded Theory: Strategies for Qualitative Research*. New York: Aldine de Gruyter.

Globe D, Schoua-Glusberg A, Paz S, Yu E, Preston-Martin S, Azen S, Varma R. (2002). Using focus groups to develop a culturally sensitive methodology for epidemiological surveys in a Latino population: findings from the Los Angeles Latino Eye Study (LALES). *Ethnicity and Disease*, 12, 259–266.

Goerman P. (2006). Adapting cognitive interview techniques for use in pretesting Spanish language survey instruments. Statistical Research Division Research Report Series, Survey Methodology #2006–2003. US Census Bureau. http://www.census.gov/srd/papers/pdf/rsm2006-03.pdf (accessed July 8, 2013).

Goerman PL, Caspar RA. (2010). Managing the cognitive pretesting of multilingual survey instruments: a case study of pretesting in the U.S. Census Bureau Bilingual Spanish/English Questionnaire. In: JA Harkness, M Braun, B Edwards, T Johnson, L Lyberg, PPh. Mohler, B-E Pennell, TW Smith (eds) *Survey Methods in Multinational, Multiregional, and Multicultural Contexts*. Hoboken, NJ: John Wiley & Sons.

Goerman P, Fernández L, Quiroz R. (2013). Translation of US Educational Level Survey Questions into Spanish: Is Adaptation the Solution? *Presented at the American Association for Public Opinion Research meeting*, May 16–19, Boston, MA.

Goffman E. (1959). *The Presentation of Self in Everyday Life*. New York: Anchor Books.

Goffman E. (1974). *Frame Analysis: An Essay on the Organization of Experience*. New York: Harper Colophon.

Gotham KF, Staples WG. (1996). Narrative analysis and the new historical sociology. *The Sociological Quarterly*, 37(3), 481–501.

Greene JC, Caracelli VJ, Graham WD. (1989). Toward a conceptual framework for mixed-method evaluation designs. *Educational Evaluation and Policy Analysis*, 11(3), 255–274.

Groves R, Couper M. (1998). *Nonresponse in Household Interview Surveys*. New York: John Wiley & Sons.

Groves RM, Lyberg L. (2011). Total survey error: past, present, and future. *Public Opinion Quarterly*, 74(5), 849–879. http://poq.oxfordjournals.org/cgi/doi/10.1093/poq/nfq065 (accessed March 15, 2013).

Groves RM, Fowler F, Couper M, Lepkowski J, Singer E, Tourangeau R. (2009). *Survey Methodology*. Hoboken, NJ: John Wiley & Sons.

Guillemin F, Bombardier C, Beaton D. (1993). Cross-cultural adaptation of health-related quality of life measures: literature review and proposed guidelines. *Journal of Clinical Epidemiology*, 46(12), 1417–1432.

Harkness JA. (2003). Questionnaire translation. In: JA Harkness, F van de Vijver, PPh. Mohler (eds) *Cross-Cultural Survey Methods*. Hoboken, NJ: John Wiley & Sons. pp. 35–56.

Harkness JA, Schoua-Glusberg A. (1998). Questionnaires in translation. In: JA Harkness (ed) *Cross-Cultural Survey Equivalence* [ZUMA-Nachrichten Spezial 3]. Mannheim, Germany: ZUMA. pp. 87–126.

Harkness JA, Van de Vijver FJR, Johnson TP. (2003). Questionnaire design in comparative research. In: JA Harkness, F van de Vijver, PPh Mohler (eds) *Cross-Cultural Survey Methods*. Hoboken, NJ: John Wiley & Sons. pp. 19–34.

Harkness JA, Pennell B-E, Schoua-Glusberg A. (2004). Questionnaire translation and assessment. In: S Presser, J Rothgeb, M Couper, J Lessler, E Martin, J Martin, E Singer (eds) *Methods for Testing and Evaluating Survey Questionnaires*. Hoboken, NJ: John Wiley & Sons. pp. 453–473.

Harkness JA, Pennell B-E, Villar A, Gebler N, Aguilar-Gaxiola S, Bilgen I. (2008). Translation procedures and translation assessment in the world mental health survey initiative. In: RC Kessler, UT Bedirhan (eds) *The World Health Organization Mental Health Survey*. New York: Cambridge University Press. pp. 91–143.

Harkness JA, Villar A, Edwards B. (2010). Translation, adaptation, and design. In: JA Harkness, M Braun, B Edwards, T Johnson, L Lyberg, PPh. Mohler, B-E Pennell, TW Smith (eds) *Survey Methods in Multinational, Multicultural and Multiregional Contexts*. Hoboken, NJ: John Wiley & Sons. pp. 117–140.

Harwell D. (2013). Violence Against Children Survey (VACS) Cognitive Interview Study, Malawi: Final Report. Hyattsville, MD: Centers for Disease Control and Prevention, National Center for Health Statistics.

Hiles D. (2008). Transparency. In: LM Givens (ed) *The Sage Encyclopedia of Qualitative Research Methods*. Thousand Oaks, CA: Sage Publications.

Hiles D, Cermak I. (2008). Narrative psychology. In: C Willig, W Stainton-Rogers (eds) *The Sage Handbook of Qualitative Research Psychology*. Thousand Oaks, CA: Sage Publications.

Johnson RB, Christensen L. (2008). *Educational Research Quantitative, Qualitative, and Mixed Approaches*. Thousand Oaks, CA: Sage Publications.

Johnson RB, Onwuegbuzie AJ. (2004). Mixed methods research: a research paradigm whose time has come. *Educational Researcher*, 33(7), 14–26.

Kane MT. (2013). Validation as a pragmatic, scientific activity. *Journal of Educational Measurement*, 50, 115–122.

Krosnick JA. (2011). Experiments for evaluating survey questions. In: K Miller, J Madans, G Willis, A Maitland (eds) *Question Evaluation Methods*. New York: John Wiley & Sons.

Krosnick JA, Alwin DF. (1987). An evaluation of a cognitive theory of response order effects in survey measurement. *Public Opinion Quarterly*, 51, 201–219.

Krosnick JA, Presser S. (2010). Questionnaire design. In: JD Wright, PV Marsden (eds) *Handbook of Survey Research*, 2nd edition. West Yorkshire, UK: Emerald Group.

Levi-Strauss C. ([1962] 1966). *The Savage Mind*. Chicago, IL: University of Chicago Press.

Lincoln Y, Guba E. (1985). *Naturalistic Inquiry*. Newbury Park, CA: Sage Publications.

Lofland J, Lofland L. (1995). *Analyzing Social Settings: A Guide to Qualitative Observation and Analysis*. Belmont, CA: Wadsworth.

Madans J, Miller K, Maitland A, Willis G, eds (2011). *Question Evaluation Methods*. Hoboken, NJ: John Wiley & Sons.

Massey M. (2014). *Analysis of Cross-National Cognitive Interview Testing of Child Disability Questions*. Hyattsville, MD: Centers for Disease Control and Prevention, National Center for Health Statistics.

Massey M, Chepp V, Creamer L, Ramirez F. (2013). Testing of the Questions on Second Hand Smoke for the National Health and Nutrition Examination Survey (NHANES): Results of Interviews Conducted 09/18/12–01/30/13. Hyattsville, MD: Centers for Disease Control and Prevention, National Center for Health Statistics.

McCall GJ. (2003). The me and the not-me: positive and negative poles of identity. In: PJ Burke, TJ Owens, RT Sherpe, PA Thoits (eds) *Advances in Identity Theory and Research*. New York: Kluwer Academic/Plenum Publishers. pp. 11–25.

Mead GH. (1934). *Mind, Self, and Society*. Chicago, IL: University of Chicago Press.

Mensch, B, Kandel DB. (1988). Underreporting of Substance Use In a National Longitudinal Youth Cohort. *Public Opinion Quarterly*, 52(1), 100–124.

Merriam SB. (2002). Assessing and evaluating qualitative research. In: SB Merriam (ed). *Qualitative Research in Practice*. San Francisco, CA: Jossey-Bass.

Merriam S. (2009). *Qualitative Research: A Guide to Design and Implementation*. San Francisco, CA: Jossey-Bass.

Miles MB, Huberman AM. (1994). *Qualitative Data Analysis: An Expanded Sourcebook*, 2nd edition. Thousand Oaks, CA: Sage Publications.

Miller K. (2008). Results of the Comparative Cognitive Test Workgroup Budapest initiative module. http://wwwn.cdc.gov/QBANK/report%5CMiller_NCHS_2008Budapest Report.pdf (accessed April 17, 2014).

Miller K. (2011). Cognitive interviewing. In: J Madans, K Miller, A Maitland, G Willis (eds) *Question Evaluation Methods: Contributing to the Science of Data Quality*. Hoboken, NJ: John Wiley & Sons.

Miller K, Maitland A. (2010). A mixed-method approach for measurement construction for cross-national studies. Joint Statistical Meetings, Vancouver, CA.

Miller K, Ryan JM. (2011). Design, Development and Testing of the NHIS Sexual Identity Question. http://wwwn.cdc.gov/qbank/report/Miller_NCHS_2011_NHIS Sexual Identity.pdf (accessed April 17, 2014).

Miller TR, Spray JA. (1993). Logistic discriminant function analysis for DIF identification of polytomously scored items. *Journal of Educational Measurement*, 30, 107–122.

Millsap RE, Everson HT. (1993). Methodology review: statistical approaches for assessing measurement bias. *Applied Psychological Measurement*, 17, 297–334.

Office of Management and Budget. (2006). Standards and Guidelines for Statistical Surveys. http://www.whitehouse.gov/sites/default/files/omb/inforeg/statpolicy/standards_stat_surveys.pdf (accessed January 22, 2014).

Olick JK, Vinitzky-Seroussi V, Levy D. (2011). *The Collective Memory Reader*. New York: Oxford University Press.

Organisation for Economic Co-operation and Development. (2006). PISA 2006 database. Retrieved July 20, 2010, from http://pisa2006.acer.edu.au/downloads.php.

Organisation for Economic Co-operation and Development. (2009). *Programme for International Student Assessment (PISA)*. Paris, France: Author.

Organisation for Economic Co-operation and Development. (2011). *Programme for the International Assessment of Adult Competencies (PIAAC)*. Paris, France: Author.

Padilla JL, Benítez I. (2014). Validity evidence based on response processes. *Psicothema*, 26, 136–144.

Padilla JL, Benitez I, Castillo M. (2013). Obtaining validity by cognitive interviewing to interpret psychometric results. *Methodology*, 9, 113–122.

Pan Y, de la Puente M(2005). Census Bureau guidelines for the translation of data collection instruments and supporting materials: documentation on how the guideline was developed (Research Report Series #2005–06). Statistical Research Division, Washington, DC: US Bureau of the Census.

Pan Y, Landreth A, Park H, Hinsdale-Shouse M, Schoua-Glusberg A. (2010). Cognitive interviewing in non-english languages: a cross-cultural perspective. In: J. Harkness, M. Braun, B. Edwards, T. Johnson, L. Lyberg, PPh. Mohler, B-E. Pennell, T. Smith (eds) *Survey Methods in Multinational, Multicultural and Multiregional Contexts*. Hoboken, NJ: John Wiley & Sons, pp. 91–115.

Penfield RD. (2005). DIFAS: differential item functioning analysis system. *Applied Psychological Measurement*, 29, 150–151.

Penfield RD, Alvarez K, Lee O. (2009). Using a taxonomy of differential step functioning to improve the interpretation of DIF in polytomous items: an illustration. *Applied Measurement in Education*, 22, 61–78.

Plummer K. (1981). Homosexual categories: some research problems in the labeling perspective of homosexuality. In: K Plummer (ed), *The Making of the Modern Homosexual*. London: Hutchinson. pp. 53–75.

Plummer K. (1995). *Telling Sexual Stories: Power, Change, and Social Worlds*. New York: Routledge.

Presser S, Rothgeb J, Couper M, Lessler J, Martin E, Martin J, Signer E. (2004). *Methods for Testing and Evaluating Survey Questionnaires*. Hoboken, NJ: John Wiley & Sons. Q-Bank, http://wwwn.cdc.gov/QBANK/Home.aspx (accessed April 17, 2014).

Reed JS. (1989). On narrative and sociology. *Social Forces*, 68(1), 1–14.

Reeve, B. (2011). Applying item response theory for questionnaire evaluation. In: J Madans, K Miller, A Maitland, G Willis (eds). *Question Evaluation Methods*. Hoboken, NJ: John Wiley & Sons. pp. 105–125.

Reichardt CS, Rallis SF. (1994). Qualitative and quantitative inquiries are not incompatible: a call for a new partnership. *New Directions for Program Evaluation*, 61, 85–91.

Richardson L. (1990). Narrative and Sociology. *Journal of Contemporary Ethnography*, 19, 116–135.

Ridolfo H, Schoua-Glusberg A. (2009). Testing of NHANES A-CASI Reactions to Race Questions Results of Interviews Conducted November 2008–February 2009. Hyattsville, MD: Centers for Disease Control and Prevention, National Center for Health Statistics.

Ridolfo H, Miller K, Maitland A. (2012). Measuring sexual identity using survey questionnaires: how valid are our measures? *Sexuality Research and Social Policy*, 9(2), 113–124.

Rubin H, Rubin I. (1995). *Qualitative Interviewing: The Art of Hearing Data*. Thousand Oaks, CA: Sage Publications.

Rubin H, Rubin I. (2012). Sharing results. In: *Qualitative Interviewing: The Art of Hearing Data*, 3rd edition. Thousand Oaks, CA: Sage Publications. pp. 213–232.

Rust PC. (1993). 'Coming out' in the age of social constructionism: sexual identity formation among lesbian and bisexual women. *Gender & Society*, 7(1), 50–77.

Schoua-Glusberg A. (2005). Hispanic Adult Tobacco Survey: Cognitive Testing Report. Hyattsville, MD: Centers for Disease Control and Prevention, National Center for Health Statistics.

Schoua-Glusberg, A, Carter, W, Martinez-Picazo, E. (2008). Measuring Education among Latin American Immigrants in the U.S.: A Qualitative Examination of Question Formulation and Error. Presented at the 3MC Conference, June 25–29. Berlin, Germany.

Schwarz N. (2007). Cognitive aspects of survey methodology. *Applied Cognitive Psychology*, 21, 277–287.

Sha M, Park H. (2013). Adapting the translation of the American Community Survey in Chinese and Korean. *Presented at the American Association for Public Opinion Research meeting*, May 16–19, Boston, Massachusetts.

Sheatsley P. (1983). Questionnaire construction and item writing. In: P Rossie, J Wright, A Anderson (eds) *Handbook of Survey Research*. New York: Academic Press.

Singer, E, Frankel MR, Glassman, MB. (1983). The Effect of Interviewer Characteristics and Expectations on Response. *Public Opinion Quarterly*, 47(1), 68–83.

Sireci S, Padilla JL. (2014). Validating assessments: introduction to the special section. *Psicothema*, 26, 97–99.

Small ML. (2011). How to conduct a mixed methods study: recent trends in a rapidly growing literature. *Annual Review of Sociology*, 37, 57–86.

Smilkstein G. (1978). The Family APGAR: a proposal for family function test and its use by physicians. *Journal of Family Practice*, 6, 1231–1239.

Smith T. (1995). Little things matter: a sampler of how differences in questionnaire format can affect survey response. http://www.amstat.org/sections/srms/Proceedings/papers/1995_182.pdf (accessed April 17, 2014).

Spanish Ministry of Health and Consume. (2006). *Encuesta Nacional de Salud de España 2006* [National Health Survey in Spain 2006]. Madrid, Spain: Ministerio de Sanidad y Consumo.

Strauss A, Corbin J. (1990). *Basics of Qualitative Research: Grounded Theory Procedures and Techniques*. Newbury Park, CA: Sage Publications.

Sudman S. (1983). Applied sampling. In: P Rossi, J Wright, A Anderson (eds) *Handbook of Survey Research*. San Diego, CA: Academic Press.

Sudman S, Bradburn N. (1983). *Asking Questions*. San Francisco, CA: Jossey-Bass.

Sudman S, Bradburn M, Schwarz N. (1996). *Thinking About Answers: The Application of Cognitive Processes to Survey Methodology*. San Francisco, CA: Jossey-Bass Publishers.

Survey Research Center. (2010). Guidelines for Best Practice in Cross-Cultural Surveys. Ann Arbor, MI: Survey Research Center, Institute for Social Research, University of Michigan. Retrieved August, 20, 2013, from http://www.ccsg.isr.umich.edu/.

Suter WN. (2012). Qualitative data, analysis, and design. In: *An Introduction to Educational Research: A Critical Thinking Approach*, 2nd edition. Thousand Oaks, CA: Sage Publications.

Tashakkori A, Creswell JW. (2007). Exploring the nature of research questions in mixed methods research. *Journal of Mixed Methods Research*, 1, 207–211.

Tashakkori A, Teddlie C. (1998). *Mixed methodology: combining qualitative and quantitative approaches*. Thousand Oaks, CA: Sage Publications.

Tourangeau R. (1984). Cognitive sciences and survey methods. In: T Jabine, M Straf, J Tanur, R Tourangeau (eds) *Cognitive Aspects of Survey Methodology: Building a Bridge Between Disciplines*, Washington, DC: National Academy Press. pp. 73–100.

Tourangeau R, Rips L, Rasinki K. (2000). *The Psychology of Survey Response*. Cambridge, MA: Cambridge University Press.

Tucker C. (1997). Methodological issues surrounding the application of cognitive psychology in survey research. *Bulletin de Methodologie Sociologique*, 55 (June), 67–92.

Van de Vijver FJR, Leung K. (1997). *Methods and Data Analysis for Cross-Cultural Research*. Newbury Park, CA: Sage Publications.

Van Maanen J. (1988). *Tales of the Field: On Writing Ethnography*. Chicago, IL: Chicago University Press.

Van Widenfelt BM, Treffers PDA, de Beurs E, Siebelink BM, Koudijs E. (2005). Translation and cross-cultural adaptation of assessment instruments used in psychological research with children and families. *Clinical Child and Family Psychology Review*, 8(2), 135–147.

Villar A, Bautista R, Palmer D, Abundis F, Harkness JA. (2006). Understanding Effects of Different English to Spanish Translations: A Case Study of ISSP Attitudinal Questions. *Paper presented at the Annual Conference of the American Association for Public Opinion Research*, May 18–21, Montreal, Canada.

Weiss RS. (1994). Writing the report. In: *Learning from Strangers: The Art and Method of Qualitative Interview Studies*. New York: The Free Press. pp. 183–206.

Werner O, Campbell D. (1970). Translating, working through interpreters, and the problem of decentering. In: R Naroll, R Cohen (eds) *Handbook of Cultural Anthropology*. New York: American Museum of Natural History.

White H. (1987). *The Content of the Form: Narrative Discourse and Historical Representation*. Baltimore, MD: Johns Hopkins University Press.

Widdop S, Fitzgerald R, Gatrell L. (2011). European Social Survey Round 6 Cognitive Pretesting Report. London: Centre for Comparative Social Surveys. http://wwwn.cdc.gov/QBANK/report%5CESS_2011_Round%206%20Cognitive%20Interview%20Pretesting.pdf (accessed April 17, 2014).

Willis G. (1994). Cognitive interviewing and questionnaire design: a training manual. *Cognitive Methods Staff Working Paper No.7*. Hyattsville, MD: Centers for Disease Control and Prevention, National Center for Health Statistics.

Willis G. (2004). Cognitive interviewing revisited: a useful technique in theory? In: S Presser, J Rothgeb, M Couper, J Lessler, E Martin, J Martin, E Singer (eds) *Methods for Testing and Evaluating Survey Questionnaires*. New York: John Wiley & Sons.

Willis G. (2005). *Cognitive Interviewing: A Tool for Improving Questionnaire Design*. Thousand Oaks, CA: Sage Publications.

Willis G, Miller K. (2011). Cross-cultural cognitive interviewing: seeking comparability and enhancing understanding. *Field Methods*, 23(4), 331–341.

Willson S. (2006). *Cognitive Interviewing Evaluation of the 2007 Complementary and Alternative Medicine Module for the National Health Interview Survey*. Hyattsville, MD: Centers for Disease Control and Prevention, National Center for Health Statistics.

Willson S. (2012). *Division of Health Care Statistics (DHCS) 2012 National Ambulatory Medical Care Survey Asthma Management Supplement Study: Results of cognitive interviews conducted August–September, 2011*. Hyattsville, MD: Centers for Disease Control and Prevention, National Center for Health Statistics.

Willson S. (2013). *Cognitive Interview Evaluation of the Federal Statistical System Trust Monitoring Survey, Round 1: Results of interviews conducted in October, 2011*. Hyattsville, MD: Centers for Disease Control and Prevention, National Center for Health Statistics.

Willson S, Gray C. (2010). *Cognitive Interview Evaluation of the 2010 Youth Traffic Safety Questionnaire: Results of interviews conducted August–September, 2009*. Hyattsville, MD: Centers for Disease Control and Prevention, National Center for Health Statistics.

Wilson T, LaFleur S, Anderson DE. (1996). The validity and consequences of verbal reports about attitudes. In: N Schwarz, S Sudman (eds) *Answering Questions: Methodology for Determining Cognitive Process in Survey Research*. San Francisco, CA: Jossey-Bass.

Yin RK. (2009). Reporting case studies: how and what to compose. In: *Case Study Research: Design and Methods*, 4th edition. Thousand Oaks, CA: Sage Publications. pp. 165–192.

Zerubavel E. (1997). *Social Mindscapes: An Invitation to Cognitive Sociology*. Cambridge, MA: Harvard University Press.

Zieky M. (1993). Practical questions in the use of DIF statistics in test development. In: PW Holland, H Wainer (eds) *Differential Item Functioning*. Hillsdale, NJ: Lawrence Erlbaum. pp. 337–347.

Zumbo BD. (2009). Validity as contextualized and pragmatic explanation, and its implications for validation practice. In: RW Lissitz (ed) *The Concept of Validity: Revisions, New Directions and Applications*. Charlotte, NC: Information Age. pp. 65–82.

INDEX

Cognitive Interviewing Methodology, First Edition.
Edited by Kristen Miller, Stephanie Willson, Valerie Chepp, and José-Luis Padilla.

WILEY SERIES IN SURVEY METHODOLOGY

Established in Part by WALTER A. SHEWHART AND SAMUEL S. WILKS

The ***Wiley Series in Survey Methodology*** covers topics of current research and practical interests in survey methodology and sampling. While the emphasis is on application, theoretical discussion is encouraged when it supports a broader understanding of the subject matter.

The authors are leading academics and researchers in survey methodology and sampling. The readership includes professionals in, and students of, the fields of applied statistics, biostatistics, public policy, and government and corporate enterprises.

ALWIN · Margins of Error: A Study of Reliability in Survey Measurement
BETHLEHEM · Applied Survey Methods: A Statistical Perspective
BETHLEHEM, COBBEN, and SCHOUTEN · Handbook of Nonresponse in Household Surveys
BIEMER · Latent Class Analysis of Survey Error
*BIEMER, GROVES, LYBERG, MATHIOWETZ, and SUDMAN · Measurement Errors in Surveys
BIEMER and LYBERG · Introduction to Survey Quality
BIEMER · Latent Class Analysis of Survey Error
BRADBURN, SUDMAN, and WANSINK · Asking Questions: The Definitive Guide to Questionnaire Design—For Market Research, Political Polls, and Social Health Questionnaires, *Revised Edition*
BRAVERMAN and SLATER · Advances in Survey Research: New Directions for Evaluation, No. 70
CALLEGARO, BAKER, BETHLEHEM, GÖRITZ, KROSNICK, and LAVRAKAS (editors) · Online Panel Research: A Data Quality Perspective
CHAMBERS and SKINNER (editors) · Analysis of Survey Data
COCHRAN · Sampling Techniques, *Third Edition*
CONRAD and SCHOBER · Envisioning the Survey Interview of the Future
COUPER, BAKER, BETHLEHEM, CLARK, MARTIN, NICHOLLS, and O'REILLY (editors) · Computer Assisted Survey Information Collection
COX, BINDER, CHINNAPPA, CHRISTIANSON, COLLEDGE, and KOTT (editors) · Business Survey Methods
*DEMING · Sample Design in Business Research
DILLMAN · Mail and Internet Surveys: The Tailored Design Method
FULLER · Sampling Statistics
GROVES and COUPER · Nonresponse in Household Interview Surveys
GROVES · Survey Errors and Survey Costs
GROVES, DILLMAN, ELTINGE, and LITTLE · Survey Nonresponse
GROVES, BIEMER, LYBERG, MASSEY, NICHOLLS, and WAKSBERG · Telephone Survey Methodology
GROVES, FOWLER, COUPER, LEPKOWSKI, SINGER, and TOURANGEAU · Survey Methodology, *Second Edition*
*HANSEN, HURWITZ, and MADOW · Sample Survey Methods and Theory, Volume 1: Methods and Applications
*HANSEN, HURWITZ, and MADOW · Sample Survey Methods and Theory, Volume II: Theory

*Now available in a lower priced paperback edition in the Wiley Classics Library.

HARKNESS, BRAUN, EDWARDS, JOHNSON, LYBERG, MOHLER, PENNELL, and SMITH (editors) · Survey Methods in Multinational, Multiregional, and Multicultural Contexts

HARKNESS, VAN DE VIJVER, and MOHLER (editors) · Cross-Cultural Survey Methods

HUNDEPOOL, DOMINGO-FERRER, FRANCONI, GIESSING, NORDHOLT, SPICER, and DE WOLF · Statistical Disclosure Control

KALTON and HEERINGA · Leslie Kish Selected Papers

KISH · Statistical Design for Research

*KISH · Survey Sampling

KORN and GRAUBARD · Analysis of Health Surveys

KREUTER (editor) · Improving Surveys with Paradata: Analytic Uses of Process Information

LEPKOWSKI, TUCKER, BRICK, DE LEEUW, JAPEC, LAVRAKAS, LINK, and SANGSTER (editors) · Advances in Telephone Survey Methodology

LESSLER and KALSBEEK · Nonsampling Error in Surveys

LEVY and LEMESHOW · Sampling of Populations: Methods and Applications, *Fourth Edition*

LUMLEY · Complex Surveys: A Guide to Analysis Using R

LYBERG, BIEMER, COLLINS, de LEEUW, DIPPO, SCHWARZ, TREWIN (editors) · Survey Measurement and Process Quality

LYNN · Methodology of Longitudinal Surveys

MADANS, MILLER, and MAITLAND (editors) · Question Evaluation Methods: Contributing to the Science of Data Quality

MAYNARD, HOUTKOOP-STEENSTRA, SCHAEFFER, and VAN DER ZOUWEN · Standardization and Tacit Knowledge: Interaction and Practice in the Survey Interview

MILLER, WILLSON, CHEPP, and PADILLA (editors) · *Cognitive Interviewing Methodology*

PORTER (editor) · Overcoming Survey Research Problems: New Directions for Institutional Research, No. 121

PRESSER, ROTHGEB, COUPER, LESSLER, MARTIN, MARTIN, and SINGER (editors) · Methods for Testing and Evaluating Survey Questionnaires

RAO · Small Area Estimation

REA and PARKER · Designing and Conducting Survey Research: A Comprehensive Guide, *Third Edition*

SARIS and GALLHOFER · Design, Evaluation, and Analysis of Questionnaires for Survey Research, *Second Edition*

SÄRNDAL and LUNDSTRÖM · Estimation in Surveys with Nonresponse

SCHWARZ and SUDMAN (editors) · Answering Questions: Methodology for Determining Cognitive and Communicative Processes in Survey Research

SIRKEN, HERRMANN, SCHECHTER, SCHWARZ, TANUR, and TOURANGEAU (editors) · Cognition and Survey Research

SNIJKERS, HARALDSEN, JONES, and WILLIMACK · Designing and Conducting Business Surveys

STOOP, BILLIET, KOCH and FITZGERALD · Improving Survey Response: Lessons Learned from the European Social Survey

SUDMAN, BRADBURN, and SCHWARZ · Thinking about Answers: The Application of Cognitive Processes to Survey Methodology

UMBACH (editor) · Survey Research Emerging Issues: New Directions for Institutional Research No. 127

VALLIANT, DORFMAN, and ROYALL · Finite Population Sampling and Inference: A Prediction Approach

WALLGREN and WALLGREN · Register-based Statistics: Administrative Data for Statistical Purposes, *Second Edition*

*Now available in a lower priced paperback edition in the Wiley Classics Library.

CPSIA information can be obtained at www.ICGtesting.com
Printed in the USA
BVOW11s0051180814

363071BV00001B/1/P

9 781118 383544